Paradigms and Barriers

Howard Margolis

Paradigms & Barriers

How Habits of Mind
Govern Scientific Beliefs

The University of Chicago Press
Chicago and London

Howard Margolis is professor in the Harris Graduate School of Public Policy
Studies at the University of Chicago. He is the author of *Selfishness, Altruism,
and Rationality* (1984) and *Patterns, Thinking, and Cognition* (1987), both published by the University of Chicago Press.

The University of Chicago Press, Chicago 60637
The University of Chicago Press, Ltd., London
© 1993 by The University of Chicago
All rights reserved. Published 1993
Printed in the United States of America
02 01 00 99 98 97 96 95 94 93   1 2 3 4 5
ISBN: 0-226-50522-7 (cloth)
        0-226-50523-5 (paper)

Library of Congress Cataloging-in-Publication Data

Margolis, Howard.
    Paradigms & Barriers : how habits of mind govern scientific belief /
Howard Margolis.
        p. cm.
    Includes bibliographical references and index.
    1. Science—Philosophy—History. 2. Science—History. 3. Kuhn,
Thomas S. I. Title.
Q174.8.M37   1993
501—dc20   92-44650
                                                                CIP

To Russell Hardin

When we look at living creatures ... one of the first things that strike us is that they are bundles of habits.

William James, *The Principles of Psychology*

Having for thirty years believed and taught the doctrine of phlogiston ... I for a long time felt inimical to the new system, which represented as absurd that which I hitherto regarded as sound doctrine; but his enmity ... springs only from force of habit.

Joseph Black to Lavoisier, 1791

Contents

Preface

 This is a further installment of what I once thought would be one book about persuasion and belief in the context of social choice. But now it is only the second of what will be a sequence of three. In the way I described in the preface to my *Patterns, Thinking, and Cognition,* the project proved to be more ambitious than I intended. *Patterns* barely reaches questions of social judgment. It does so mainly in its concluding chapter on Galileo's encounter with the Inquisition. That left a considerable agenda of topics that had to be worked out to open the way to what might be—what of course I hope will prove to be—a fruitful attack on some questions more centrally in the domain of politics and social choice. The pope's problem with Galileo's book was certainly a political, not intrinsically a theological and certainly not a scientific, problem. But it could be analyzed in terms of the situation between the two men, with no attention to what are really more difficult questions of what shapes public opinion, and how that influences the perceptions and constrains the choices of political actors.

 In a 1992 article, I tried to fill in the Galileo account as published in the concluding chapter of *Patterns.* Another area that needed follow-on work lay in providing stronger empirical evidence than I was able to provide in *Patterns* for the habits-of-mind ("scenario") account of illusions of judgment worked out in chapter 8. Richard Griggs (1989, 1990) provided independent replication of enough of those results to encourage me that the scenario account was on the right track. But he could not replicate enough of that work to remove some doubt that that account really got to the heart of the illusions-of-judgment problem, which in turn was enough to make it apparent that more work on my part was called for. Unless I run into further trouble, a report on experiments providing stronger results should be ready by the time the present book appears. The third book of this sequence should be ready to go to press the following year, reporting work on a salient problem in the realm of environmental politics that seemed to me exceptionally fruitful for my purpose. Its empirical focus is on the striking disparity that has sometimes appeared between expert judgment and the judgment of concerned, sophisticated nonexperts.

I am indebted for support for the early phases of this work to the National Science Foundation program in history and philosophy of science. Eventually its reviewers lost confidence in the project, about the same time that others from the same community seemed to be gaining confidence in the same work. So we will see how things turn out, now that the final result is available. But I am grateful for the support that did come, which was particularly valuable since it came before I enjoyed the security of tenure at Chicago.

As an outsider, I am of course indebted more generally to many people within history or philosophy of science who have provided information and advice, not always followed, but always useful. Among these, in one way or another, Larry Holmes, Persi Diaconis, Phil Kitcher, Steve Stigler, Sam Westfall, Al Van Helden, Tom Nickles, David Hull, Bob Richards, and Tom Kuhn have been particularly helpful. Since much of the study exploits case material from the history of astronomy, I owe a special debt to Owen Gingerich and Noel Swerdlow for many discussions of technical points, and to Swerdlow also for translation of a key passage from Hobbes (in chapter 11). And as is apparent even from my title, the work could never have been undertaken without the path-breaking provided by Kuhn, to whom I owe a personal debt as well as an immense intellectual debt.

In a more general way, I want to mention a number of people who have been supportive for years, including Jack Ruina, George Rathjens, Gene Skolnikoff, Lucien Pye, Carl Kaysen, and Tom Schelling. I owe a particular debt to Bill Kruskal, who was acting dean of the School of Public Policy at Chicago during a crucial period for me, and to Russell Hardin, who was primarily responsible for the appointment at Chicago, which more than anything else made it practical to carry through this work. And I owe the most sustained debt to my wife. As the saying goes, it is nice to be married to your best friend.

Introduction

The aim of this study is to give an account of the Kuhnian notion of "paradigm shifts" that makes sense in terms of the account of persuasion, belief, and judgment worked out in my *Patterns, Thinking, and Cognition.* A sketch of such an account was given in *Patterns* itself (chapter 9), followed by a series of applications. That is greatly elaborated here. The introduction to *Patterns*, which provides a summary of the overall argument of that earlier book, is reprinted as appendix A. Cross references generally give chapter and section, for example, *Patterns* [2.7].

As in *Patterns*, I set out blunt basic claims and try to stick to them. A fruitful research motto, at least for me, has been "simple ideas, stubbornly pursued." Of course the simple ideas become modified as they are stubbornly pursued but, in this case at least, not in the sense of adding qualifications. On the contrary, I have been led to strip away qualifications. Bertrand Russell (1956) once remarked that "the point of philosophy is to start with something so simple as to seem not worth stating, and to end with something so paradoxical that no one will believe it." I will try to move more nearly in the opposite way, starting from an idea likely to strike many readers as implausible, but hoping to persuade at least some of those readers that the idea works out too neatly to be far wrong.

The simple idea, inherited by this study from *Patterns,* is that, in terms of what is neurally encoded in brains, there is unlikely to be any essential difference between a physical habit and a habit of mind. But physical habits are easier to observe, and hence more familiar as definite things with characteristic properties. Thinking about physical habits should, on this view (or conjecture, if you wish), be a very good and convenient guide to thinking about habits of mind. And habits of mind—now picking up the central idea of *Patterns*—are really all there is to cognition.

In the opening sentences of *Patterns* I supposed that pattern-recognition is all there is to cognition. But the two points are complementary ways of saying the same thing. The pattern-recognition process requires the prompting of a pattern that "looks right" (or more often some subconscious analogue of "looks right"). In turn, together with other features of the context, this prompts a pattern of response that experience in the world, or the congealed experience of genetic endowment, has linked to that pattern. Talk of habits focuses on the linkage between patterns; talk of pattern-recognition focuses on the repertoire of patterns. But the first presumes the second, and the second would be useless for getting along in the world without the first.

These basic claims do not imply that cognition is simple, any more than the simplicity of Newton's laws means that physics is simple. But starting from something simple allows complications to arise in application without making the analysis intractable. It quickly leads to an idea that is sketched but not developed very far in *Patterns* (chapter 9). The idea is that, when we talk about a Kuhnian paradigm shift, what we are essentially talking about is a special sort of shift of habits of mind. I try to show how this leads to the far more counterintuitive claim that, ordinarily, a Kuhnian paradigm shift comes down to the breaking of some one particular habit of mind: the *barrier* of my title. The earlier chapters deal almost exclusively with developing the barrier argument. But as the study proceeds, we will be increasingly concerned with complementary aspects of the habit-of-mind view, stressing the role of endemic propensities—the genetic endowment that is prior to habit.

Of course the notion of "habits of mind" has been familiar at least since the time of Hume. But I will try to put that familiar notion to work in a more than metaphorical way, and as more than a handwave to account for otherwise puzzling stubbornness. The discussion starts by developing the argument that habits of mind can be treated as tightly parallel to physical habits, whose reality and importance for governing behavior no one doubts. In chapter 1, I try to show why there is much plausibility to the *identification* of habits of mind with physical habits, and hence to pursuing parallels between the two in an aggressive way.

Chapter 2 considers what properties we might expect of habits of mind if we pursue the parallel with physical habits. I tie the argument here in particular to the discussion of learning in chapter 7 of *Patterns*. We will be concerned with the notion of *rivalry* (what it takes to present a challenge to entrenched habits) and how rivals are ordinarily *tamed* but occasionally lead to radical restructuring of previously well-entrenched beliefs.

Chapter 3 shows how we are led to a particular account of the central feature of radical discovery (the barrier hypothesis). I try to show how the habits-of-mind claim implies the possibility of the phenomena Kuhn has captured under the notion of "paradigm shifts," and hence suggests a particular account of what is going on when we observe the stresses that characterize such shifts. The most striking claim, as I have already mentioned, is that ordinarily a particular, identifiable habit of mind is critical for the emergence and contagion of just those new ideas capable of provoking marked symptoms of Kuhnian incommensurability. The chapter develops this *barrier* view of paradigm shifts in contrast to what I take to be the (usually only tacit) assumption that what provokes the Kuhnian symptoms of radical innovation is some logically difficult *gap* dividing an established theory from its challenger. I develop a good deal of detail on the contrast between this gap view of paradigm shifts and the barrier view urged here.

The next several chapters apply these ideas to a series of famous episodes in the history of science. As this proceeds, the main focus of the argument shifts to the nature of belief formation and change in general (no longer with the exclusive focus on revolutionary episodes), which in turn ties to contemporary debates about the nature of science and the evolution of scientific ideas. In the historical material, I try to show how the barrier analysis can make sense of known details of these episodes in ways that have not been seen before. The overall coherence of the argument and its pragmatic effectiveness in concrete cases give what seems to me good grounds for taking seriously the theoretical argument about habits of mind of the earlier chapters.

The main case studies I use cover Lavoisier's overthrow of phlogiston (chapters 4 and 5), the emergence of probability (chapter 6), and the Copernican discovery (chapters 7–10), with the last providing a good deal of new material extending the Copernican account presented in *Patterns*.

The Lavoisier analysis shows how the barrier view leads to an account of this episode in which—putting the matter a little strongly—the turning point is the analysis of water in 1783, not the discovery of oxygen by (independently) Priestley, Scheele, and Lavoisier some eight years earlier. Even after suitable qualification, we are left with quite a different account from the usual one—the usual facts are seen to form quite a different pattern of events. I try to show how this account unfolds very naturally in terms of the barrier viewpoint, though it has been almost completely missed from the more usual gap viewpoint. I also try to show how, once this view of the events is seen, many customarily ignored details of the controversy make obvious sense instead of standing as anomalies that don't comfortably fit the story, and hence tend to be left out of the story.

The account in chapter 6 of the emergence of probability turns on a corollary of the barrier hypothesis. In the overwhelming majority of cases, habits of mind serve as facilitators of effective thinking, not as barriers, just as physical habits ordinarily facilitate effective performance, not impede it. But then cases should arise in which some discovery is not blocked by some well-entrenched but incompatible habit of mind, but rather by the absence of some essential facilitating habit of mind. I try to show how the emergence of probability can be interpreted as a striking example of this possibility.

I identify a particular cognitive move (what I call "inverse counting") which is now so commonplace and habitual as to be almost invisible but which appears only in the mid-seventeenth century. The emergence of probability then follows very promptly after several thousand years during which it was (logically) easily available but never grasped. I try to show how the required insight into the usefulness of inverse counting itself emerges through work on concrete problems salient for Harvey, Torricelli, and (most of all) Galileo. Their work then promoted sensitivity to "inverse" possibilities across the developing community of science, so that an idea that is easy today finally became possible for a few of the most brilliant thinkers of the mid-seventeenth century.

Chapter 7 (first of the Copernican chapters) takes a different form from the primarily historical chapters 4–6. The bulk of it is in fact a tutorial on the relations among Copernican, Ptolemaic, and Tychonic astronomy. The tutorial treatment tries to show in a particularly transparent fashion how the observational equivalence of all these systems is a simple consequence of the geometry of Ptolemaic astronomy. This apparatus is put to work in chapters 8–10 to clarify several subsidiary details of the Copernican discovery: for example (chapter 9) to show how a few numbers Copernicus had noted on a scratch page bound with his personal copy of the Alphonsine tables provide some direct evidence that he in fact proceeded to his discovery in the way conjectured as part of the habits-of-mind analysis presented in *Patterns*.

The tutorial material of chapter 7 also provides a point of departure for the more general discussion of scientific belief of the concluding chapters. These chapters begin by applying the tutorial material to bring out the very striking extent to which what is seen as persuasive evidence and argument may be shared widely across communities: in this case of pre-Copernican astronomy, across enormous spans of time and large differences in culture. The material is also intended to illustrate the value of using what we pragmatically know about the world (contra much concern about whiggishness) to make sense of what was going on in earlier science. The examples show how various puzzles about the development of *Ptolemaic* astronomy can be resolved—and indeed, as far as I can see, could only be resolved—by exploiting our contemporary knowledge that the world is not Ptolemaic but *Copernican*.

In contrast to claims that using the fruits of science to make sense of what was going on in earlier science is somehow unsound, I try to show by concrete example how a bit of whiggishness in fact is the key to making sense of the greatest achievements of early science. Chapter 8 gives a highly sympathetic account of why it would have been extremely difficult for Ptolemy *both* to develop the magnificently effective system he did *and* to see that the system could be turned inside out to yield the system that waited another fourteen centuries to be discovered. The Lavoisier chapters try to bring out the brilliance of Stahl's phlogiston theory. For both, on the habits-of-mind account, it is the brilliance of the earlier achievement that is crucial for understanding why the subsequent emergence of alternative ideas, though logically not difficult at all, nevertheless proved to be stubbornly difficult.

In chapter 11, I turn to an analysis of the controversy between Hobbes and Boyle, and show how the habit-of-mind view leads to an entirely different analysis of this episode than that presented by Shapin and Schaffer in their widely admired constructivist study. The critical methodological points turn on the appropriateness of taking seriously the possibility that there are such things as good arguments and good evidence that are essentially independent of a particular social context.

The study then provides a habits-of-mind account of the emergence of a novel sense of experiment in the first decades of the seventeenth century. The argument is closely parallel to that on the emergence of a novel sense of counting in the account of probability in chapter 6. The notion of a novel sense of experiment as the critical development of the Scientific Revolution is a common one. But the articulation of just what was involved in it seems to be new.

The general point to be drawn from these examples is that, although we get our clearest picture of what is going on as belief changes in cases where a barrier is in operation (so that otherwise inexplicable cognitive responses alert us to look for what is accounting for those responses), the role of habits of mind is pervasive and ordinarily helpful. The most deeply entrenched cognitive propensities can be expected to hold across wide spans of time and culture because these most deeply entrenched cognitive propensities will be just those tuned to essentially universal experience in the world. The barrier cases show us how striking habits of mind can be in blocking what later come to seem irresistible intuitions. But that evidence of the force of habits of mind also assures us that habits of mind and the underlying cognitive propensities that shape habits of mind must also powerfully constrain intuition in ordinary conditions, where there does not seem to be any rationally puzzling oddity to be explained. These endemic propensities are always around, stubbornly pushing things their way, so that it is not surprising that sooner or later things in fact do move their way. I try to show how such propensities, interacting with the characteristic features of scientific communities, can be expected to severely constrain what beliefs can be sustained over time by such a community. The argument yields, I think, a strong case for what we might call "the limited relevance of relativism."

One

Habits of Mind

H abits of mind" suggests entrenched responses that ordinarily occur without conscious attention, and that even if noticed are hard to change. We are interested in the role of habits of mind in the evolution of scientific belief—and by implication, though I won't say much explicitly, in the evolution of belief in general. A remark that shifting from one paradigm to another requires retraining intuitions, or reshaping habits of mind, is commonplace. But I want to make that remark more precise. Its consequences are neither trivial nor always immediately apparent.

Although the habits-of-mind idea is not of itself at all novel, its usual usage has been only as an explanation of last resort. If the logic of a claim seems clear but belief doesn't appear, talk is prompted about cognitive inertia, conceptual difficulties, and so on, including habits of mind. Here that is reversed. On the argument, I am committed always to giving an explanation primarily in terms of habits of mind. Logic enters the discussion, and indeed commonly plays a large role in it. But it is the logic that is derivative, in the way worked out in detail in *Patterns* (chapter 5). All activity controlled by the brain is treated here as the cuing of patterns (of imagery, thought, motion, and so on).

The cues can be external (perceptions) or internal (priming or inhibiting effects of recently activated patterns), or most commonly both. Learning can then be characterized as the modification of patterns in the repertoire, and as the training of new linkages among patterns.

Talk of habits ordinarily refers to patterned responses that we can see (even if the person who has the habit cannot). But beyond quantum effects, observable responses are not plausibly different in kind from unobservable ones. What is observable is ordinarily a function of what we are equipped and alerted to observe, not of any deep property of the thing at issue, so that observability per se is not the feature we want to focus on. We want to focus on another feature of ordinary language usage: plasticity. Habits can be changed, but not easily. Later in the study we will also be much concerned with a category of cognitive features that appear to be genetically entrenched (instinctive) and cannot be changed: what I will call endemic propensities. For the learning and occasional deflection of habits must be shaped by underlying endemic propensities.

The operation of a physical habit involves triggering a particular pattern, which prompts the sending of signals to the muscles appropriate for carrying out the pattern, given the current disposition of the body and its relation to the environment. So the triggering of a physical habit results in a characteristic pattern of physical movement. Triggering habits of mind results in a characteristic pattern of intuition. On a Darwinian view, we could expect continuity across these forms of patterned response, and readily observed empirical material strengthens that presumption (*Patterns,* chapters 2 and 3).

I argue that we might as well suppose an essential identity (at the level of what is embodied in the brain) between physical habits and habits of mind. With respect to such matters as how it is acquired, how it works, how it can sometimes be changed, I will suppose that there is no essential difference between a physical habit and a habit of mind other than the one obvious and gross difference that a physical habit involves patterns of movement and a habit of mind yields patterns of thought, intuitions, ideas, images, intentions. With that qualification, the conjecture to be developed here is that differences between entrenched patterns governing movement and those governing intuitions, ideas, and images are comparable to whatever differences could be expected within the uncontroversially mental categories (images, intuitions, and so on).

A parallel as tight as I will be sketching between physical habits and habits of mind is not a standard view within psychology today. This may mainly reflect the near disappearance of the notion of "habit" from psychology, except in the border area of sports psychology. For some decades, behaviorism ruled out any notion that invited talk of mental states. But the notion of habit is closely connected with ideas about what a person does with conscious attention versus what usually occurs without that. The characteristic focus of behaviorism was on learning by animals, which came especially easily just because with animals such mentalistic notions as "trying to learn," "looking for what's going wrong," and so on are easily preempted. There is hardly room in a behaviorist framework to talk of such things, nor of a category of responses to stimuli—habits—which was ordinarily unconscious but which involved the possibility that its consequences could sometimes be brought to conscious attention, noticed to be giving bad results, and subjected to conscious effort at change by working at producing an alternative response.

With the cognitive turn, consciousness is once again a respectable thing to talk about. But it is not something anyone can yet do much with. It is not something that anyone has a clue to implementing as a computer program. Consciousness becomes a notion now often discussed in a philosophical way, but at the periphery of research. Things were entirely different for writers at the start of the twentieth century. Psychologists then took their subject to *be* the science of consciousness, and habit then is salient. In his long chapter on habit, William James (1890) seems to have taken for granted the identification between physical habits and habits of mind. He slips midway through the chapter from discussing physical habits to discussing habits of mind with no indication that he feels he has changed the subject in any way. So although identifying habits of mind with physical habits is not common today, it is also not at all a new idea, or perhaps even a controversial one. It has been neglected rather than resisted.[1]

1.2

Consider an illustration of how habit works that dates back half a century (Bernstein 1933). Suppose you are asked to write something on the blackboard of a lecture hall. You will be able to do that without hesitation, and your handwriting will have the characteristic features it ordinarily exhibits. This would be true even if you had never written on a blackboard before. Yet writing on the blackboard (since the scale will easily be enlarged ten times or more) obviously involves an entirely different set of detailed motions than normal writing, with large roles played by muscles that would not even be involved in ordinary handwriting. You could also write, although clumsily, but still inescapably with characteristic features of your own handwriting, with muscles that have never been used at all for writing, such as with a pencil between your toes or in your mouth.

Evidence of this sort demonstrates that the physical habits involved in handwriting and, by easy extension with similar examples, physical habits in general, cannot be patterns of particular muscle movements, since the exact muscle movements required to produce the externally observable result must vary greatly from one context to another. Hence what is in the head as the neural embodiment of these patterns can only be patterns somehow neurally embodied, not particular sets of commands to the muscles, and certainly not something locally "in" the muscles that ordinarily carry out the pattern. Not only is the locus of handwriting or violin-playing habits in the brain (not the hand, as might be naively supposed), but what is in the brain is not a sequence of instructions for muscle movements but a pattern that can be applied to guide muscles that ordinarily have nothing to do with writing or violin-playing.

The further argument, then, is that certainly the simplest, and I think also the most plausible, hypothesis about the relation between such a pattern used to guide muscles and a corresponding pattern that lets us *recognize* the pattern once physically present in the world— what relates the pattern that makes the physical habit and the pattern that a person can recognize or even produce as a mental image—is *identity*. They appear to be the same thing, though an unambiguous case (from study of lesions, etc.) where the two lead to conflicting results would challenge this unqualified claim.[2]

Even when the result is being produced with the usual muscles, habitual physical responses cannot possibly be carried out by a fixed set of neural signals (a particular set of instructions to the muscles). As Bernstein pointed out, writing a word on a vertical pad involves moving up and down in the gravitational field, but writing with your arm resting on a table does not. So signals to the muscles producing sequences of contractions of particular force and timing that would yield the intended output in one case would yield gibberish in the other. What controls the motions, then, must be something like (nonconscious) images of motions, which guide signals to the muscles appropriate to this particular occasion. A person can sometimes visualize such patterns and mentally rehearse the motions, sometimes even notice what has been done wrong from reviewing this mental imagery.[3] We can expect that, as we learn more, we will find a subtly controlled anchor-and-adjust process.

1.3

Consider some well-entrenched response that we would not ordinarily think of as a habit, such as your immediate recognition of the
meaning of a familiar word. Habit has the connotation that in favorable
circumstances a person may come to preempt or alter the usual automatic response. So is recognition of a word a habit that could be
changed? The answer is certainly yes. Nearly all words convey slightly
different meanings in different contexts, and many words convey entirely
unconnected meanings (such as *bank* in the money context and *bank*
in the river context). The relation between cues (indications of context)
and responses to cues is subject to change, most conspicuously so in
acquiring the specialized meaning of a term in some technical context.
In certain contexts a mathematician comes to no longer see the once
automatic meaning of *imaginary* or an economist *demand* or *utility* or
a pool shooter *run,* and so on, but instead just as automatically sees a
new meaning that has become habitual in those contexts.

Note here a fundamental point for the whole discussion: a critical change in habits need not mean eradication of some established
habit. Much more often, in fact, contexts remain in which the old habit
survives untouched, a conspicuous example being ordinary life intuitions and talk about the motion of the sun. If we allow the notion of
habit to be extended to the meaning habitually attached to words, we
easily point to examples of that sort, as just illustrated.

1.4

In *Patterns* [2.4]* I suggested that things we know how to do
could be partitioned into instincts, habits, and judgment, where habits
are built from instincts (and simpler habits), and judgment turns on the
ability to sometimes explicitly consider more than one alternative (so
that a choice is consciously made among alternative responses to a
stimulus). All cognition, on this view, consists of linked sequences of
pattern-recognition ("*P*-cognitions"), most of which is inaccessible to
introspection. Cases of explicit judgment are some tiny fraction of all
the choices we make. Even the most self-aware cognitive processes (the
special form of judgment we call calculation) consist substantially of
unconscious steps. And even steps that could be interrupted and examined are very often difficult to treat in that way, like trying to stop

---

*Here and elsewhere, *Patterns* [x.y] refers to chapter x, section y of Margolis 1987.

yourself in the middle of a turn of phrase or the middle of a snatch of melody. That can sometimes be done, but a person has to be motivated to make a special effort to manage it.

Now notice that many things that every normal human being comes to know how to do (such as sitting or walking [physical capabilities], or speaking [a mental capability]) are never simply a matter of genetics. No one is born programmed to walk with a military bearing, any more than she is born programmed to speak Chinese rather than English, or to speak with a particular accent. But as Chomsky argues, the sort of languages human babies readily learn, like the gaits they readily learn, must be constrained by innate propensities (that is, instincts relevant to acquiring language or mobility). Similarly, while there are variants across cultures in things like usual postures, facial expressions, and so on, nevertheless basic features of all these things are common across cultures. We will be concerned with a parallel distinction in characterizing what it is that changes in the course of a paradigm shift: it will be the equivalent of a gait or accent, while more fundamental entrenched responses—what I call endemic propensities—not only do not change, but cannot change. That we can expect to find well-marked endemic propensities yields the possibility that we can say something useful about the conditions under which entrenched habits are sometimes successfully challenged.

1.5

Knowing a language involves knowing how to produce a very elaborate set of muscular operations (comparable to the motions of an expert violinist in controlling his instrument) that actually produce the spoken language. The conspicuous difference between the cases of speech and violin-playing is that we are born with a propensity to learn to speak, so that ordinarily (without explicit training) innate physical propensities are tuned up by experience to produce the required entrenched habits of performance. Hence we acquire this skill without conscious effort. But learning the violin, which is not so close to what is in our genes, requires a tremendous effort. A person who knows a language also has acquired the ability to translate patterns of mechanical motion in the ear into things our minds understand. On the other hand, "body language," which we are more inclined to think of as governed by physical habits, must involve something of the propensities for communicative behavior of language in general. Sign language is an intermediate case, only recently recognized to be much closer to spoken language than to the nonsyntactic signals of body language.

We appear often to use the very same patterns both in movement and in thought. A person can mentally "review" a golf or tennis shot and sometimes "see" what he has done wrong. Or a mathematically inclined person can visualize how a curve would change as parameters are changed, and also can draw the alternative curve on a blackboard. It does not seem likely that the purely mental language you think with could be different from the physically expressed language you communicate with, nor that the pattern that guides merely thinking about a mathematical curve is really different from the pattern that guides the hand in actually drawing the curve.

For these cases the line is fuzzy between patterns we know in a mental sense and patterns we know in the physical sense of knowing how to move our muscles to get a certain result. A habitual response will rarely be wholly eradicated, as a person who once learns a language or how to play the violin will be only "rusty," not incapable if asked to perform after many years of disuse. Experience can alter the salience of various responses in some context, come to produce novel responses in that context, and come to ordinarily inhibit what had been a well-entrenched response. But since the original habit will not be wholly eradicated, it may still occasionally appear, though if noticed it may now be seen as a mistake.

1.6

What else would follow from the conflation of physical habits with habits of mind urged here? A good deal of the response I want to make is best worked out in the context of the concrete cases that will make up the largest part of the study. But some further general implications can be set out as starting points for that discussion.

1. Acquiring physical habits, as everyone knows from their own experience, requires practice (repeated experience). No one can just decide to keep her eye on the ball, however much she is convinced that is the thing to do. The least controversial point about parallels between physical and mental habits is that the need for practice holds for both. It takes practice to become fluent in a language or a new piece of mathematics or a new concept. And it is particularly hard to learn a new habit when that habit competes with (requires displacing) one already routinely cued in the relevant context. Of particular importance, we will see, is the rival habit that is not only well-entrenched but also entangled in multiple ways in the practice of the actor.

2. Since habitual responses are ordinarily evoked without conscious attention, we all have habits we are unaware of, and even for

habits we are aware of, we easily fail to notice occasions on which they are activated. For physical habits, no one doubts that. But if someone says we have a certain physical habit, it is usually easy to persuade us that it is so. Often we can immediately see it ourselves, once attention is drawn to it. Or we can see it with the help of a mirror or videotape. Or a friend or coach can see it. None of this is available to facilitate our seeing habits of mind. If we could only observe the results of a habitual physical act (the golf ball is struck, but we don't see the swing), we could not notice anything at all of the detailed bits of well-learned movement that produce the result.

Suppose, however, we had a series of still photographs taken during the swing. If they were sufficiently close together, we could then see the habit in action (if really close, we would have a movie of the swing). Even a few shots might be enough to let us tease out some inferences about governing habits. The analogue of the snapshots for habits of mind are the conscious glimpses of intuitions taking shape that a person can report, though these are certainly only a small fraction of all that goes on to support thinking. Other clues for ferreting out the governing habits of mind would come from contrasting responses to other contexts resembling in some way the context we are trying to understand, or from a similar pattern of response in some contrasting context.

So although a habit of mind cannot be directly observed, the consequences of a habit of mind are things that can be seen, in the sense that we can see how a person responds to various contexts: what intuitions she reports, how he responds to arguments and pieces of evidence, and so on—including the reportable "snapshots" of conscious intuition that occur on the way to a judgment.

A person with the appropriate training sees a certain squiggly line not as the squiggly line an untrained person sees but as (say) a seismograph tracing, while another person (with different training) sees it as a electrocardiogram. If it is pointed out that that particular tracing in fact is an electrocardiogram, our seismologist might easily say, "Well, it does look odd for a seismogram, but from force of habit I saw it that way." So it is not hard to notice that our subject has a *habitual* tendency to see a tightly sinusoidal curve as a seismogram. If we watched her closely, we would be able to compile a considerable list of further habits of mind likely to be characteristic of anyone who is an expert seismologist. Similarly, by watching how members of some expert group respond to situations, what they take for granted, what they immediately question, we can learn something about the habits of mind that are characteristic of that group, which in turn can tell us something about why a particular belief was hard or easy to produce.

3. In the vast majority of situations a person's habits of mind (like his physical habits) will serve him well. It is just the possession of a peculiar set of habits of mind that makes a person an expert seismologist, as it is a collection of well-entrenched and highly interacting physical habits that facilitates effortlessly driving a car or playing the piano, or which in general makes for skilled performance in any activity (Dreyfus and Dreyfus 1986). But since a person is ordinarily unaware of habits, she will also ordinarily be unaware of when a habit is not serving her well. In particular, since a person—even if alerted to a habit—cannot just "turn it off" to see if things would go better, changing habits is clumsy and requires time. It may be more difficult still to see that there is in fact a habit at work and that it is a habit that may not be working well, so that there is an occasion to make an effort to change.

We will be much interested in the special circumstances in which an individual might become aware that something has gone wrong, which is the way a person might be prompted to work her way out of some entrenched habit of mind. As a simple example, the seismologist who sees a tightly sinusoidal curve in the (for her) habitual way might be jarred by sharply conflicting evidence (for example, by reading the labelling of the figure) into realizing that her professional experience has on this occasion led her astray. From an instance this stark, this seismologist (more readily than another, equally competent) might on another occasion be open to the possibility that what she had taken for a seismogram with somewhat odd features might be in fact not an odd seismogram but a typical something else.

Naturally, there is a spectrum of such possibilities. At one extreme the anomalous evidence or argument—the sources of a *rival* intuition[4]—is stark, like the golf shot that is a bad slice, or the seismologist's confrontation with a caption that tells her that what she is looking at is no seismogram. At the other extreme are minor perturbations that even the most self-critical person would treat as only an annoyance to be put out of mind. As with physical habits, a person will not be prompted to the work required to challenge entrenched habits of mind unless confronted with disturbing results. For a given challenge, the prospects for effective disruption of an entrenched habit—for disturbance sufficient to prompt the sustained effort required to change a habit—will be contingent on the balance between the strength of the habit and that of conflicting habits comparably well-entrenched, which might prompt rival intuitions in this context (*Patterns* [7.6]). Somehow, anomalies must be encountered that prompt rival intuitions in a sufficiently striking and stubborn way that the consequence can be an eventual change in habits, not just a taming of the anomaly. In principle, even deeply entrenched habits can always be challenged if there is evidence and argument that yields intuitions that conflict with those prompted by the habit. In practice, that does not happen easily, sometimes even when evidence and argument that in hindsight look compelling are available. In some ways changing a habit of mind will be more difficult than changing a physical habit.

Use of a physical habit will be observable (to others, though often not to the actor himself), while a habit of mind cannot be directly observed, and in fact need have no external output at all. The snapshots mentioned earlier do not show us the habit, but only clues to what the habit might be. When I see an argument or a work of art or whatever in a valuative way, that valuation will be shaped by my habits of mind. My intuition may be that it looks good (not bad), right (not wrong). But behind that evaluation lies my habitual sense of what looks appropriate or reasonable in this context. The reportable intuition (it looks good or looks bad) gives no direct indication about what habits of mind were involved in prompting that response.

So compared to physical habits, habits of mind must be hard to detect and identify, which means that the problem of changing a habit of mind will be complicated by our ordinary lack of any conscious sense that a habitual response is in play, and by the absence of introspective access to what the habit might be, even if we are motivated to want to know. Striking evidence on this point is available from experiments on "illusions of judgments."

4. Habits that develop spontaneously with experience in the world must be harder to notice than habits that are explicitly learned, hence that a person knows about even if routine performance prompts no explicit notice of what is going on. But among the class of tacitly acquired habits are many that develop in connection with overt drill and training but are not explicitly taught. Another category is the "imported" habit that a person perhaps knows he has (because he was explicitly trained to acquire it or because it has come to explicit notice in some other way) but that is being tacitly used in contexts *different* from those in which it was acquired.

A person learning to ski explicitly acquires the habit of putting his weight on the downhill ski, but he picks up many other habits that are not explicitly taught. Since a person is especially likely to be unaware of tacitly acquired habits, they are especially susceptible to being imported to contexts where they may have perverse consequences. But habits of mind are essentially never explicitly taught, and hence almost never explicitly noticed. When habits of mind do change, we can expect that even more than for physical habits the person does not see what he is doing differently. He discovers his intuitions have changed—what had looked obviously wrong now looks okay—but he cannot put his finger on why that change should have occurred. Since it is current intuitions that govern judgment, she is puzzled by why she once saw the item as wrong, not by why she now sees it as right. On the argument here, that is the basis of Kuhnian conversion experiences.

5. Polanyi's (1974) "tacit knowledge" must reflect habits of mind. Again, as with habits in general, in the vast majority of situations this serves a person well. But there is no *cognitive* dichotomy between good habits and habits that can profitably be challenged. Any good habit could turn out to be perverse in some context. And a habit that a person has no way to notice will be particularly hard to challenge.

In particular, when everyone in a community shares a habit, it ordinarily becomes invisible, for what everyone does no one easily notices. We commonly pick up physical habits in ways that reveal an unconscious imitation of styles of doing this or that typical of the community in which we acquire those habits. Particularly obvious examples concern speech habits (accents, turns of phrase, and so on).

We cannot easily notice our own accent. Our accent, customary tones of voice, turns of phrase, and so on are examples of habits so deeply and generally entrenched across a culture that they become almost invisible. Ordinarily no one thinks of them or notices them. This will occur for physical habits as well as for habits of mind. But such social effects must be stronger for habits of mind. If I try to hit a tennis ball or make a ski turn in an unusual style, I may look odd to bystanders. But there is no need for the bystanders to understand what I am trying to do differently, however, nor for the move to look sensible to bystanders, in order for it to become apparent whether it is effective. But new ideas profit from discussion, and if the possibility I am exploring looks absurd or incomprehensible to others, I may be cut off from exchanges that might be fruitful, or helpful in finding support, or merely congenial even if functionally irrelevant. And while new ideas might sometimes yield immediately striking empirical power, it is usually only after sustained work that an innovator has arguments and evidence that could convince outsiders. So there may be a sustained period through which what the innovator is doing does not make sense to others.

Put another way: since habits of mind are intimately tied to communication, the people we communicate with freely—all the more so on some scientific or otherwise technical issue that exploits concepts and arguments far from ordinary discourse—are people who share specialized habits of mind with us, and hence, as we say, see things the way we do. "Talking past each other" yields the converse case. All this makes habits of mind even more likely to be socially shared than (say) styles in a physical activity, like skiing, and harder to escape. Incompatible habits of mind block communication, easily evoke resentment and distaste and frustration, all of which would tend to reinforce a natural propensity toward coordination of habits across individuals and make breaking with socially shared habits of mind harder than breaking with socially shared physical habits.

6. Yet persuasion, belief, discovery, and the like can start only in some individual brain. However large a role social processes play in shaping habits of mind, often in giving an issue salience, and always in turning what must start as individual idiosyncrasy into something "everyone knows," at the point of emergence discovery always is an individual story.[5] Radical discovery in particular—discovery that at first easily prompts incredulity, confusion, or even revulsion—must turn on something interestingly atypical about the habits of mind of the individual discoverer, or something interestingly atypical about the experience of that individual, or commonly some mix of both.

Each cognitive response (seeing things in a certain way on a certain occasion) is intrinsically an individual thing—as much so as having a headache or drinking a gulp of water. Since no two individuals are identical either in their makeup or their experience, no two individuals will share exactly the same cognitive repertoire. Rather, habits of mind, hence intuitions, must certainly vary across individuals, even though, on the whole, we can expect the socially shared component of even a radical discoverer's habits of mind to be far more extensive than the idiosyncratic component. We want to understand how what initially are slight variations in habits of mind and experience sometimes have remarkable consequences.

That is not always or even ordinarily a challenging task. A discovery may be nearly inescapable given the situation, so that the fruitful part of the analysis will in fact concern social features that effectively assure a usual response from every individual, not the details of individual response. Somehow a particular individual sees something that others in the community have not seen. But often the discoverer is only reacting to some novel experience in the way that anyone in the community is likely to react. If the experience is of a sort that can be readily shared with others, contagion of the new idea is also unproblematical.

But while there are cases in which it is easy to see, given some novel development (such as invention of a more sensitive observing instrument), why a discovery was made when and where it was, there are also cases where prevailing social features cannot account for the discovery, since most individuals with the same information as the discoverer were blind to the discovery and commonly at first resisted it even when it was brought to their attention. What we want to understand in such cases is how that individual (or cluster of individuals[6]) came to make the discovery. One reason for attending closely to these exceptional cases—what we would call the cases of revolutionary discovery—is that in such cases we get our best chance to gain insight into the fundamental processes that govern routine as well as exceptional discovery. Barring miracles, what is going on in radical discovery must be a particular case of ordinary cognitive processes. But in the context of radical discovery, we have a chance to see the emergence of a new way of seeing that is in some way starkly unexpected, and under conditions that generate intense scrutiny. In particular, it is under just such conditions that we have our best chance of getting useful glimpses of habits of mind, since here the consequences are stark and striking. As the study proceeds, we will be led to consider how analysis of these revolutionary cases might help us understand more routine cases.

7. If you swing at a golf ball one way and my swing is different, that in no way interferes with our enjoying a round together. Problems turning on coordination of physical habits are familiar (for example, dancing), but they arise in special contexts in which coordination is an explicit value. When we are in such a situation, the habitual movements that need to be changed are visible. All this is reversed for habits of mind, as already discussed. And for intuitions, coordination is not ordinarily valued for its own sake, however much we are all tacitly influenced by the propensity to coordinate. Explicit coordination of habits of mind is even less plausible, since they are invisible. So although coordination of habits of mind commonly occurs across a community, it is a strictly tacit coordination (as with our common sense of nuance in language), which ordinarily attracts no explicit attention (it is taken for granted) and is certainly not managed by any explicit attempt to secure coordination. The closest thing to a counterexample is entirely compatible with the argument here: it is the case where one party sees itself in an inferior state with respect to the intuitions at issue, for example, the disciple who is explicitly trying to learn to see things as the master does.

Outside of that special case, we all like to think that we believe what it is sensible to believe, not what someone else claiming superior judgment believes. So with respect to habits of mind that make it easy to believe some things, and hard to believe or even make sense of others, individuals ordinarily will not be motivated to seek coordination. An individual with a radical idea ordinarily wants to persuade others to see things his way, not to adjust his beliefs so that he can be coordinated with the others.[7]

So with respect to habits of mind, situations must arise in which intuitions conflict, and we cannot see easily or at all why they conflict. In addition, neither side sees any reason to try to see things as their adversaries do: rather, the puzzle for them lies in why their adversaries cannot see how weak their case is. Hence difficulties arise that have no strong analogue for the case of physical habits. Arguments that seem powerful to one side seem unimportant to the other. What looks like striking insight to one side looks like perverse illusion to the other. Often, the parties simply see the world differently, in some way that is not directly observable (since habits of mind are invisible and operate below the level of consciousness) and is tied to no reasonable definition of where their interest lies. It is convenient to have a special term for such conflicts. What I take to be essentially this phenomenon has been most clearly identified and articulated by Kuhn, and his term is the one I will use: incommensurability.

Two

Paradigms

On the whole, the social way in which science is studied and practiced and communicated must ordinarily make it difficult for individuals to come to beliefs that violate the strong intuitions of everyone around them.[1] Nevertheless, radical change of belief sometimes takes hold in a community, and that must start with some individual initiative, as the evolution of species can start only with individuals who vary in what turns out to be some significant way from others in the collectivity. This suggests a habits-of-mind account of what it is that shifts when a paradigm shifts. On this account, an element of Kuhn's argument that has often been taken to be particularly open to doubt—incommensurability—becomes the central notion, as indeed it has become increasingly central in Kuhn's own work.

When scientific ideas change smoothly, we assimilate the episode to the "normal" end of Kuhn's normal-to-revolutionary spectrum. When change occurs in a wrenching way that provides conspicuous evidence of incommensurability—later in this chapter I will be more explicit about that evidence—we have cases of Kuhn's revolutionary science. Since it is easiest to see what is going on in the stark cases, we will mainly be concerned with cases near the revolutionary pole.

Incommensurability in the "talking past each other" sense

stressed at the end of chapter 1 certainly occurs outside science. It is conspicuous especially in politics. But science provides peculiarly favorable conditions for understanding the phenomenon. Here the parties involved are characteristically highly expert and largely share the same expert judgments, which makes it possible to focus narrowly on points of divergence to an extent that usually could not be approached in a political context. In a controversy with a substantial political character, it may be hard to judge how far "talking past each other" might be merely tactical and fundamentally governed by conflicting motivations born of conflicting interests. Conviction is tied not only to the logic of the arguments but to interests and alliances and the pressure to act, whether the question is ripe for a confident resolution or not.

But while there is some political element in every controversy, for controversies that mark major turning points in the history of science it is hard to point to cases where anything plausibly outweighs the interest each actor shares in not wanting to look stupid to the next generation of graduate students.[2] Scientific controversies are characteristically about what the world is like, and however naive philosophers or sociologists may judge that, it is taken for granted that there is ordinarily a right, or at least a better, answer to such questions: and an answer that sooner or later will be known.

2.2

Consider next a position I want to reject. Think for a moment of the relation between habits of mind and paradigms as something like the relation between physical habits and games. We could think of people operating within a paradigm as developing certain characteristic habits of mind that fit with and ordinarily facilitate their work, as someone who plays tennis or squash develops physical habits that facilitate play in those games. Theories and descriptions of equipment and procedures would be the analogue of rules and descriptions of the equipment and layouts for the games.

Changes in habits of mind (if this view were sound) would go along with a paradigm shift, as some physical habits would change if a person switched from squash to tennis. But for paradigms as for sports a change in habits of itself could not change the activity. Nor would it make any sense to say that habits of mind are constitutive of a paradigm. That would be like saying that the physical habits characteristic of people expert in squash or tennis are constitutive of the games themselves.

But on the view I will develop, habits *are* constitutive of paradigms. To put the point in the most extreme way: shared habits of mind are the only *essential* constituents tying together a community in the way that makes talk of sharing a paradigm fruitful. Talk of a paradigm without particular habits of mind (I argue) is like talk of a square without a perimeter. I don't really mean to deny that shared beliefs, as well as shared habits of mind, are important. But as will be seen, after all qualifications, it remains the case that (on the argument developed here) the essential component of a Kuhnian paradigm is an intrinsically invisible (though not undetectable) component, habits of mind.

To the objection that it sounds odd to say that habits of mind are constitutive of a paradigm, it seems to me a reasonable first response (though obviously not a reasonable final response) to point out that it is now almost thirty years since publication of Kuhn's *Structure of Scientific Revolutions,* yet debate is as unsettled as ever on the nature of paradigms, which suggests that some new way of thinking about the issue is worth trying.[3]

Does the claim here imply that a person with certain habits of mind *but without any knowledge of the theories that ordinarily would be considered at the heart of a paradigm* could share the same paradigm with the community characterized by those habits but with the knowledge? This seems and indeed is absurd. But not, on the view here, because it is logically absurd. Rather, entrenchment in certain habits grows out of intense, prolonged experience with certain kinds of activity, as entrenchment in the physical habits that make an expert tennis player or violinist grows out of intense experience with those activities. So as a practical matter, but not as a matter of logical necessity, we would not find the central habits of mind other than in the context of the rest of what Kuhn has labelled the "disciplinary matrix," any more than we would find the special physical habits of a violinist in anyone who was not in fact a violinist.

Hence, on the view here, the central puzzle for understanding what binds together a certain community (making communication easy within the community and making it hard to communicate with a rival community when that appears) would be identification of habits of mind which tacitly guide critical intuitions within that community in ways that would not come easily or seem reasonable to someone who is not a member of that community and who therefore lacks the intense experience with seeing things in particular ways that are characteristic of members of that community.

A paradigm *shift* then is a special sort of change in habits of mind. In particular, we are interested in just those cases where the shift is in some way essential for emergence of a new idea in science. But the features we find for this special case, we can expect, will have something in common with other cases of shifts in habits of mind in entirely different areas, such as politics or the arts.

But contrary to the analogy considered earlier, the relations of habits of mind characteristic of a scientific community to the prevailing theories and procedures of that community are essentially different from the relation between the habits characteristic of expert play of a game (even a mental game, like chess) and the rules of the game. Theories and habits of mind interact far more strongly than their analogues in the games context. Working with a theory as it evolves promotes the development of altered networks of habits of mind.[4] At the same time, whatever habits of mind are currently in force facilitate some extensions or variants of theories and serve as barriers to others, as will be discussed in concrete contexts later. But development of the theory is also shaped by the way the world is. So the way the world is constrains the way theory develops, which creates pressure for reshaping habits in ways that fit comfortably with what theory has become.

In limited ways, something like this point applies also to games. The kind of games fans like to watch and players like to play affects the rules of sports, and habits of play will change to accommodate the new rules. But the interaction between habit and theories occurs in science to an extent that has no serious parallel in the relation between physical habits specialized for some game and the rules of the game. The distinction arises in the following way. It makes no sense to say that physical habits of tennis players constitute tennis, and it would also make no sense to say that, for example, the habits of a Ptolemaic astronomer constitute Ptolemaic astronomy. But for the sport, in addition it would also make no sense to say that habits constitute the *community* of players of that sport, since a person with a radically different style of play could unproblematically remain a member of the community.[5] For a paradigm, that could not really happen. Kepler could play the game of Ptolemaic astronomy better than any actual Ptolemaic astronomer alive when he wrote his *Astronomia Nova.* But it was plain—and not only because he said so in very blunt ways—that he was not a member of that community, but someone who displayed Ptolemaic techniques merely to show how clumsy they were compared to what he advocated as the more sensible way to do astronomy. Someone who learned Ptolemaic astronomy from Kepler's treatment would not learn to see things in a way that made irresistible sense to Ptolemaic astronomers for so many centuries before Copernicus, but in a way that made it irresistible to see the Ptolemaic way as wrong.

The fundamental difference between paradigms and games arises because the core activity in science is not playing the game as it is, but learning how to change the game (research). This leads to amended habits, which, once in place, govern intuitions in the ordinarily tacit way that physical habits guide movements, without conscious attention or even awareness that such a habit exists. Dramatic shifts—the kind of change that we and the participants themselves perceive as radical—occur when there is no smooth path from one set of habits to the next. Rather, something creates a marked sense of cognitive discontinuity in the first audience for the new ideas, eventually yielding (when successful) the Kuhnian sense of conversion among those convinced, and of being swept aside by something that does not really make sense among those who are not. But as will come up repeatedly when we begin to look at concrete cases, episodes that exhibit the marks of Kuhnian incommensurability do not always involve logically difficult transitions, and episodes that involve logically difficult transitions do not always exhibit Kuhnian incommensurability. The critical discontinuity is cognitive, not logical.

2.3

The argument here might seem more congenial if I spoke of "point of view" rather than habits of mind. But there is good reason not to do that. A person is ordinarily conscious of a point of view. But unless specifically and effectively prompted, a person is ordinarily unconscious of habits and indeed is to a large extent completely unaware that she has various habits. Similarly a person ordinarily can try a different point of view and certainly understands what is being asked when someone proposes a look at things from another point of view. But we can't try out a different habit, and the claim that a person has a habit is something that usually needs to be backed up by much more than an appeal to introspection. So while the habits-of-mind view pressed here overlaps some of what is meant by the more commonplace comment that people within a paradigm share a point of view, it is not at all merely another way of stating such a claim. As will be seen, the Copernican case provides particularly sharp examples where a person is highly expert in a formal theory but not operating in the paradigm at all: she can take the Ptolemaic point of view, but she is not at all entrenched to some critical Ptolemaic habits of mind; and we can exhibit striking consequences of that.

2.4

Since habits of mind ordinarily are unnoticed, how do we ferret them out? As already discussed in chapter 1, the key lies in looking for anomalous responses. Providing a complete description of habits of mind characteristic of a paradigm is not a plausible task. We cannot realistically do that, but we also do not need to do it. We need only identification of those habits of minds that are critical for distinguishing the community from outsiders or rivals. Analysis of a paradigm takes shape in relation to some alternative view, even if the alternative view is only something like what has come to be called "folk theory." Pragmatically, habits (or even more deeply rooted reflexes or instincts) can be made visible only by setting them against the background of some explicit alternative. We can become aware of our own habits of walking by comparing it to some other gait, as we can become aware of our own habits of speech by comparing the way we talk to that of some other community that shares the language but has differences of accent and lingo. In our context, if we can identify an at least implicit alternative, we could try to say something about habits that would facilitate one way of doing things (for habits of mind, some sort of intuitions) and serve as a barrier to the alternative. But we need never undertake the impractical task of providing some total account of habits of mind characteristic of a paradigm.

Indeed, for the cases that are most useful for seeing how science evolves, there ordinarily is a salient paradigm challenged by some alternative. We are then looking for just those habits that could account for the incommensurability across a given pair of rivals. There is not some in-principle-unbounded scope to the enterprise. Conceivably it may prove hard to find and convincingly pin down the key items. But in the cases I have examined in any detail, it has not proved difficult to produce plausible candidates for this role.

Pursuing this requires no challenge to common usage. A paradigm is commonly taken to encompass what is found in a textbook, but also it is widely understood—and certainly so in Kuhn's account—to include what is between the lines of a textbook. On the account here, "between the lines" is given a particular meaning. What is between the lines of a textbook that is essential for the operation of the paradigm are critical habits of mind that tacitly guide key intuitions within the community. Such socially shared habits of mind facilitate communication and many other aspects of constructive work, but they also constrain what can be seen as making sense.

2.5

Given this viewpoint, it is clear that the notion of "paradigm shift" has to be understood (as indeed it is in Kuhn's own treatment) as a matter of degree. No sharp boundary can be drawn between the normal evolution of science and revolutionary shifts.[6] Even the most "normal" discovery, if we look at things in sufficiently fine detail, upsets some expectation encouraged by prevailing habits of mind. The element of unexpectedness is just what gives the perception of novelty. There is always some cognitive story to be told about how a discovery came to be made, and how it came to be accepted by others, and why the transition occurred when and where it did. But for the vast majority of discoveries the stories would be quite trivial, and except as an exercise no one would bother to ferret them out. Because we use the term "discovery" in an honorific sense, a routine discovery in fact would not even be referred to as a discovery, just "Smith found" or (a bit more impressive) "Smith was the first to find that . . ."

At the other extreme, episodes where the signs of Kuhnian incommensurability are well-marked give us cases that are not only of wide general interest, but that yield special insight into the way science works, including how normal science works, and more broadly yet, into the way that persuasion and belief formation work in general. Features that are characteristic of every instance of persuasion, belief, and judgment can here be seen in the large, so that details that cannot be clearly discerned ordinarily may now be plainly on view.[7]

2.6

We would have a tricky problem if we had to sharply distinguish the "matter of degree" that puts the transition between two viewpoints nearer the revolutionary rather than the normal pole of the spectrum of possibilities. But since we are interested here mainly in striking cases, the surface manifestations that indicate the presence or absence of the kind of cognitive discontinuity that we want to study will be stark.

The clearest cases—hence the cases I will usually focus on—occur when all the information required for a discovery was clearly in hand for a long time prior to the discovery. Then on the face of things, there was something decidedly revolutionary about the discovery. (Or why was it not seen earlier?) For example, the Copernican discovery easily qualifies as revolutionary in these terms. As will be discussed in more detail later, Copernicus's heliocentric argument requires no information not available to any reader of Ptolemy, and hence to every significant Western or Middle Eastern astronomer for almost fourteen centuries before Copernicus finally saw what was (logically) readily available the whole time.[8] We will see that a similar claim holds also for the emergence of probability and a milder but still striking and instructive example is provided by the overthrow of phlogiston.

Striking delay is neither the only nor a necessary (although it is a characteristic) symptom of a revolutionary discovery. If a discovery is made, and evidence and argument that eventually will look convincing are in hand, but the relevant specialist community—or at least some important segment of that community—somehow cannot see the sense of the argument, that again is prima facie evidence for a revolutionary episode, here focused on contagion (possibly but not necessarily in addition to the discovery difficulty already introduced). Again the Copernican case provides a striking example. Copernicus opens his book by conceding that he knows his central idea will strike readers as absurd. And although his book was studied closely by astronomers for its technical information and generated sufficiently lasting demand to earn a new printing eighteen years after its publication in 1643, it is clear that in the short run Copernicus failed in his attempt to overcome the incredulity he anticipated from his initial readers (Gingerich and Westman 1989). Astronomers were studying the book despite its heliocentric claim, not because of it.

It is not until the 1580s (forty years after the Copernican publication) that we see evidence that astronomers were beginning to see the heliocentric argument as possibly making sense. So sheer span of time—how long it takes to go from a point where a discovery is clearly available to when it is actually seen, how long it takes before knowledgeable readers can see the force of the argument supporting the discovery once it has been made—provides the least ambiguous indication of a cognitively revolutionary episode. In the Copernican case we find both. But sometimes—in particular when inferences from the claim yield startling empirical success—things can move quickly, and we are aware of the cognitive difficulty (as with quantum mechanics) only because the participants talk about it. We would like an account that lets us understand how these variants arise. Finally, we want that account also to make sense of the cases in which a discovery is clearly revolutionary in the sense of marking some extremely important turn in the development of a science, but shows no sign of cognitive difficulty. The almost routine acceptance within biology of Crick and Watson's structure for DNA illustrates this case.[9]

In general, then, the history of an episode can show us when there was something strikingly difficult (or not so) about the emergence or contagion of an idea. But the surface criteria do not tell us what *made* a discovery revolutionary rather than normal. Taking note of these criteria only sets the stage for the main inquiry, which concerns characterizing and teasing out the consequences of the habits of mind that (on the view here) account for the cognitive difficulties.

Three

Barriers

The most natural view of why a few new ideas provoke marked symptoms of incommensurability [1.6], but many others do not, lies in some notion of a logical or conceptual distance between the new idea and what preceded it. We can then think of a revolutionary paradigm shift as a shift across an intrinsically large space between new and old theories or practices or concepts. Call that the gap view of paradigm shifts.

Alternatively, we can consider the quasi-Darwinian possibility that new ideas that provoke the symptoms of a Kuhnian revolutionary episode are just those where there happens to be an important conflict between new ideas and ordinarily unnoticed habits of mind that have developed in the practice of the older ideas. Then symptoms of incommensurability might be clear whether or not there is some exceptionally difficult logical or conceptual discontinuity. Of course I have been leading up to a defense of that view.

A converse possibility also arises, as will be discussed in detail in chapter 6 on the emergence of probability: an idea may be long delayed because its discovery becomes practical only after some novel habit of mind has developed out of other work. This is the case I will characterize as turning on a missing habit of mind, not on a conflict with entrenched habits of mind. If one sort of case occurs, the other must be expected also. In the case of probability, and typical of such cases, the new idea does not overthrow an existing paradigm. It opens up an area that until this idea arrived had remained unexplored.

But even in this converse case of missing habits, a new idea will always involve some conflict with some existing habits of mind, since it is just the violation of expectations that fit existing habits of mind that gives us the sense that something is new. To be persuaded, someone must find it easier to believe the new idea than not to believe it. What looks wrong about rejecting the new idea must more than offset what looks wrong about accepting it. Eventually (chapter 10) I will work out a more formal statement of this in terms of endemic propensities that govern our balancing risks of accepting a bad idea (errors of the first kind) versus the risks of neglecting a good idea (errors of the second kind). This balancing process governs responses to argument and evidence whether the new idea is a radical one with large consequences or a perfectly routine adjustment of a mistaken belief, as with the seismologist example [1.6].

Sometimes accepting the new idea—given the evidence and argument and other circumstances that encourage that—is unambiguously easier than not doing so, or the converse. But it is also common that the person goes through at least a momentary, and sometimes a prolonged, period in which there is hesitation. The argument looks good, so it looks wrong to reject the belief; but the belief itself still looks wrong. The individual is in the "paradox" state of the belief matrix of *Patterns* [5.8].

A more extreme situation is when the new idea is seen as incoherent, absurd, or perverse, yet some time later even people who were once making such judgments, or still later their students, are left puzzled about why such an obviously interesting idea was at first seen as making no sense. It is this last sort of situation—the cases that show strong symptoms of incommensurability—that we particularly want to study, starting with consideration of how such situations might arise.

On the gap view, the logical distance separating a prevailing view and some conflicting new proposal may be comfortably narrow, so that a person can just step across it. But wide gaps require a leap. Very wide gaps require an exceptionally good jumper and exceptionally favorable conditions for the leap to be feasible. Wider gaps still are beyond the reach of any plausible human leap. If the new idea is reachable at all, it will be only later in the history of the science, when theory and evidence are in place that do not require so forbidding a leap. On this view, once we know the prevailing theories before and after a shift, a judgment about the size of the logical gap explains why a revolutionary transition (a leap) was required to get across in some cases, but only a normal science evolution from earlier ideas in others.[1]

The phenomena of delay can pretty easily be accommodated to this view of things. For example, there could be delay in contagion of an idea once discovered, because at first it appears there is only about as much to be said for it as against it. Once the case is strong (if indeed it becomes strong), there would be little further delay unless the gap is large, as a person would not delay in hopping across a small puddle.[2]

The gap view sits comfortably with the "games" analogue discussed at the outset of chapter 2. Theories, procedures, and apparatus (on this gap view) are the essential constituents of Kuhnian paradigms. Habits of mind, while no one doubts that they exist, play no *essential* role. Whether a particular transition is revolutionary or not depends on how different the rival theories are, how extensive a remapping is required to go from one conceptual network to another, and so on. From the gap view, there may not be much in Kuhn's notion of paradigms beyond a single word for the complex of theories, equipment, and procedures as described in a textbook. What is between the lines is only further detail, not something (like habits of mind) that is intrinsically tacit. Talk of incommensurability is likely to seem just puzzling.[3]

But on the barrier alternative, the critical problem for a revolutionary shift in thinking is not in fact some intrinsically difficult logical or conceptual gap that needs to be leaped (though that might be present). It is the robustness of the habits of mind that block the path to the new idea, relative to the habits of mind that help the new idea, given the evidence and argument available to support it. Suppose that (as I will argue in a moment) episodes that clearly show Kuhnian symptoms of a revolutionary development characteristically involve a particular habit of mind that needs to be overcome. Call that the barrier. Ordinarily, on this view, anyone who can get over the barrier can get across the gap. The critical puzzle is not to judge the size of the gap. Rather, the critical focus is within the gap, where the primary question concerns identifying the critical barrier within that gap.

Hence when we see a shift that shows the Kuhnian symptoms (delays in seeing a possibility that is logically available, incomprehension or indifference among its first audience to arguments that eventually come to seem irresistible), we expect to be able to identify a particularly important cognitive barrier and to be able to give an account of how it came to be breached (or for what I have called the "missing habit" case, how a novel habit arose that finally put the idea in reach).

We are prepared to find that even when the symptoms of delay or incomprehension (or often but not necessarily both) are striking, nevertheless the logical steps necessary to get from the old theory to the new might be quite undemanding relative to the technical skills required to work in the field at all. As the study proceeds, I try to show this is true for the transition from Ptolemaic to Copernican mathematical astronomy, and even more transparently for the overthrow of phlogiston and for the emergence of probability.

Of course habits of mind far more often block ideas that would not only look worthless or perverse when first encountered, but would in fact be worthless or perverse. But once in a while the converse might hold. The occasional appearance of strong barrier effects is inevitable on the logic of the habits-of-mind argument. Although rare, we can expect that such episodes might play a large role in the history of science, as rare favorable mutations play a large role in Darwinian evolution.

3.2

Consider a habit of mind that would tend to block a certain intuition that (with hindsight) we know would prove powerful. Occasionally, the relevant habit could be highly robust, in the sense that it is so deeply entrenched and entangled in expert practice that escaping it is difficult even after some logical or empirical anomaly directly challenging that habit had come on the scene. Robustness in this sense would turn on the habit's being not only deeply entrenched (used routinely and unhesitatingly in dealing with core situations of the prevailing view) but also entangled in multiple ways with such practice. As the study proceeds, I will try to make these notions (entrenched, entangled) increasingly clear through concrete examples.

A habit of mind that is both highly robust and critical for the emergence of a new idea I will call a *barrier* for that case. The much stronger claim I am leading to is that characteristically, in a revolutionary episode, we should be able to identify a *unique* barrier, in the sense that a person who somehow gets beyond that barrier is almost certain to go on to make the discovery, but not so otherwise.

From the general discussion of habits of mind (chapter 1), it follows that the barrier need not be something explicitly discussed by participants (since habits of mind are ordinarily unnoticed), nor something that would attract much attention in accounts tacitly shaped by the gap view of paradigm shifts. The "nested-spheres" habit of mind is central to the barrier account of the Copernican discovery (*Patterns,* chapters 11 and 12; chapters 7–10 here). But it is rarely so much as alluded to in accounts of the Copernican discovery.[4] That is not surprising, since the nested-spheres aspect of Ptolemaic astronomy has always been treated as a peripheral feature, which logically it is. The idea is not even mentioned in Ptolemy's *Almagest,* though it is the core idea of the later, but much briefer, *Planetary Hypothesis.*

Since the barrier account of the Copernican discovery turns on an issue (the role of nested-spheres habits of mind) that is ordinarily entirely ignored, it is easy to see a large difference between the barrier account and a gap account of that episode. Such a clear divergence between gap and barrier views is not usual. A good example of the more typical case, where it requires more detail to separate the gap and barrier views, is Kepler's use of elliptical orbits and of his equal-areas rule governing speed in place of the traditional apparatus of interacting epicycles.

On all accounts, Kepler's innovation was a radical move. But the barrier that had to be overcome to see that possibility was directly tied to the main explicit steps that would be focused on in a gap account: the long-standing commitment of astronomy to the principle of uniform circular motions (UCM). Kepler dispensed with both exact circles and uniform motion. So there is a marked gap here between Kepler's mathematical models and those that had wholly dominated astronomy, starting centuries before Ptolemy and continuing through the work of Copernicus and Tycho. Chapter 7 gives an account of the tight family relationship that links the Ptolemaic, Tychonic, and Copernican formal models despite the radical differences in cosmology. But Kepler's models are radically different. And for either a gap or a barrier view of the episode, the key step is the departure from the traditional commitment to UCM.

Even though discrimination between gap and barrier views commonly requires some care, however, marked differences will nevertheless emerge from details of an episode. For Kepler, the usual account treats the explicit commitment to UCM as a kind of rule of the game, like the requirement for classical geometric construction to use only a compass and a straightedge. Trisecting an angle but using some auxiliary device (beyond compass and straightedge) would not produce an acceptable construction, since it fails to follow the rules. The pragmatic success of Kepler's ellipses let him get away with challenging the rules. But seeing that the ellipses would work was technically demanding: many astronomers of the period could not have handled it even if they had the idea and the boldness, skill, and patience to pursue the unconventional and difficult geometric analysis required. So here is a definite and plausible gap story to be told.

But Copernicus himself defends the commitment to UCM on more compelling grounds than a general deference to time-honored practice. Presumably, so did at least the more able earlier astronomers even if their explicit defenses of the principle have not survived. Copernicus argues (1543, I.4) that only if the heavenly motions are the resultant of compounded UCM could we understand the regularity with which the complex motions of the planets repeat themselves over long cycles of time. The only alternative is to postulate variable forces in the heavens. But since a compounding of several naturally perpetual rotations of spheres accounts for observations and makes physical sense, such complications (Copernicus argues) are unreasonable.

These Copernican arguments are not implausible, and they are certainly much more than blind obedience to an arbitrary rule. It was only at the end of the sixteenth century, for someone with access to Tycho's data, that information was available that could show any advantage to a break with tradition. Yet it is obvious that something more than technical difficulty had to be overcome to do what Kepler did. Even after he had spelled out his argument in 1609, it was several decades before other astronomers followed his escape from the commitment to UCM. The point is often made that Galileo did not recognize the force of Kepler's new ideas. But apparently neither did anyone else until (in the 1630s and beyond) the striking accuracy of Kepler's Rudolphine tables made it impossible to avoid the inference that there was something right about what he was doing (Wilson 1989). Kepler himself makes it clear that his "struggle with Mars" was long and difficult. Even he did not easily escape the commitment to UCM.

As will be seen later, logically the arguments favoring UCM are weaker for a Copernican or Tychonic astronomer than they are for a Ptolemaic astronomer. There is really no puzzle about the commitment to UCM before Copernicus. But there is a puzzle about why Copernicus, Tycho, Galileo, and for quite a while Kepler should have been so tightly bound by arguments that had lost an important piece of their underpinning.

What I will try to show—when we come back to this issue in chapter 8, and with parallel analysis of other aspects of the Copernican story and of the overthrow of phlogiston—is that pursuing the habits-of-mind argument gives us something much more explicit than a vague reference to cognitive inertia and the like to resolve such puzzles. Always, the barrier arguments turn on how certain intuitions would be entrenched by a person's experience in *acting* in the world. Always, the critical habit of mind is rooted in some pervasive aspect of physical experience in the world, and then entangled in various ways in expert practice.

For UCM, for example, we all have much experience with regularly recurring events, and all the most striking of these familiar experiences (a wheel coming round, the Sun rising) appear to be the result of UCM. A person also has experience with recurring motions that are not UCM but lopsided or otherwise asymmetric, and experience with such things entrenches expectations that then are unsteady, erratic, quick to lose their momentum: in short, conspicuously not perpetual. But the conspicuous feature of the heavenly motions is that they do appear to be perpetual. Explicit arguments like those used by Copernicus help articulate and explain these habitual ways of seeing things, rooted in habits of mind tied to physical experience. The explicit arguments Copernicus deploys can make sense to us today because they are consistent with intuitions rooted in familiar experience, even though no one any longer believes in UCM.

This physical basis for intuitions about UCM then would become entangled in the experience of an expert astronomer by the great convenience of UCM for doing calculations. Uniformly rotating circles are easy to calculate with. Technically, no matter how long the span of time over which we are calculating, we only need to find the displacement of the radius modulo 360 degrees. It is then a simple matter to combine the displacement of several circles to get a prediction of where among the fixed stars the heavenly body will be. Working with Kepler's ellipses is somewhat more difficult, and very much less obvious. After the superiority of his predictions was recognized following publication of the Rudolphine tables around 1630, a prime interest in technical astronomy was in working out ways to give close approximations of Keplerian predictions using computationally more convenient Ptolemaic methods (Wilson 1989). Prior to Kepler's initiative, the raw physical experience that would entrench habits of mind that supported UCM would become entangled with the professional practice of working astronomers in using UCM over and over as a highly convenient and effective way of doing their work. So escape from UCM was—as Kepler himself makes clear—very difficult even for someone who was both gifted and an ardent Copernican.

Another example used later to develop the notion of entanglement comes from the phlogiston case, where the core habit of mind derives from the extremely familiar experience of seeing something (say a candle) burn, where we all see flames leap up and dissolve in the air and after a time the candle is gone. But seeing how this became a formidable barrier that required a Lavoisier to challenge it requires seeing how this very well entrenched sense of what happens in combustion became entangled (correctly!) in Stahl's theory with superficially unrelated phenomena characterizing the reduction of ores, respiration, and the formation of acids.

I will defer technical details until we reach the discussion of later chapters. But I hope even these preliminary remarks make it clear enough that, in speaking of a habit of mind as entrenched and entangled, I am not using words with some elusive meaning: in each case I will be pointing to explicit experience in the world, of much the same nature as the experience that accounts for entrenched physical habits.

Note also that the barrier argument as sketched so far actually implies the second, converse case I have already [3.1] characterized as turning on some missing habit. This occurs when a barrier can be detected that arises, not as the consequence of the presence of a habit of mind incompatible with a new idea, but as a consequence of the absence of a facilitating habit of mind. It must sometimes be the case that a certain idea is hard to reach without the prompting of some already entrenched habit, where that facilitating habit happens to have the character that everyday experience in the world would not create it. Rather, this habit of mind could plausibly emerge only after things had developed to a point where expert practice—highly artificial expert practice unrelated to the ordinary business of life—would encourage the evolution of that habit. A concrete example will play the central role in the analysis in chapter 6 of the emergence of probability.

3.3

If all cognition is reducible to sequences of pattern-recognition (and *Patterns* amounts to a long argument for that), then what a person can do at any particular moment is constrained by the repertoire of recognizable patterns currently available. What I can make sense of at any particular moment (the intuitions that can occur to me) are constrained by the range of patterns I can recognize and by the habitual linkages among those patterns.

New patterns are added to the repertoire by a process that involves variation from already present patterns, usually in combination with novel use of existing patterns. For any discovery we should be able to propose a discovery path with the property that (subject to the qualifications just mentioned) at any point on that path the person is only doing what she already knows how to do (seeing what she already knows how to see, understanding only what she already knows how to understand), as described in much more detail in *Patterns* (chapter 7).[5]

Even if a novel intuition is presented to me by the particular circumstances of the moment, the range of novel intuitions that can look right—look like insights in that context, rather than illusions to be promptly dismissed or forgotten—must be similarly constrained by my available repertoire and by the linkages among the patterns in that repertoire. So of the range of intuitions that might occur to me under some (perhaps odd or fleeting) circumstances, only some small fraction would be seen as more than a momentary misperception. A great many intuitions that might occur to a person in one way or another will be seen as obviously wrong, as a person routinely—ordinarily with no more than fleeting notice—dismisses slips and misperceptions every day. But among that great many profitably ignored misperceptions there will be, rarely but surely, an occasional "fruitful misperception" (*Patterns* [5.4, 6.2]), which if it could be attended to for a while (rather than immediately seen as wrong) might reveal something that would not be forgotten.

So, at least if the argument of *Patterns* about cognition as pattern-recognition is more or less right, we could expect that an intuition that yields some radically new idea, or that would be an essential step toward that idea, will occasionally be among those blocked by the prevailing habits of mind: so that the idea would be unlikely to occur to a person, and even if seen, it would look wrong. As may be seen from the examples given in *Patterns* and extended here, there is nothing terribly subtle about this claim: the habits of mind I will point to by way of illustration will be (and on the argument here, characteristically would be expected to be) essential to the ideas at issue in very straightforward ways.

Such cases would give us examples of a cognitive barrier. These become striking when the barrier remains potent even though the idea is logically clearly within reach: most obviously so when the eventual discovery, when it comes, is based on no evidence or technique that had not been available for many years before anyone was able to see its significance. For the Copernican discovery and for the emergence of probability (we will see), the necessary information had been available for upward of a thousand years.

But just to the extent that there is a remarkable delay while an idea that is logically within reach somehow remains unattended to—or a remarkable stubbornness in coming to see its strength once someone has discovered it—we would have the strong indications of incommensurability discussed in chapter 2. We would have reason to infer that some such barrier indeed is at work even if we as yet have no sense of what it might be. On the barrier argument, the occasions in which barriers play an important role can be identified because they will be just those cases where clear symptoms of incommensurability are apparent.

Hence we are motivated to look for cases where the symptoms of incommensurability are markedly apparent, since it is just those cases that should provide us with occasions where looking for barriers and considering their consequences will yield particularly striking results.

3.4

Consider the process that eventually yields the critical change in habits of mind, starting from two points that should not themselves be controversial. On the one hand, habits are rarely either quickly made or quickly broken. Much repetition is characteristically required. On the other hand, anomalies ordinarily have only a modest effective lifetime. Anomalies lose their bite: in the language I used in *Patterns* [7.7], they are ordinarily tamed in some way that yields no major disruption of entrenched habits. In the language of biology, we become habituated to them. In the phlogiston case we will take up shortly, the weight gain that accompanies calcification (rusting) seems like an irresistible indication that what is involved is the gain of something, not a loss of something. The effect had been familiar long before Priestley, Scheele, and then Lavoisier discovered that oxygen is absorbed in these processes. But by then the weight gain had been thoroughly tamed, so that it was somehow not seen—even for quite a while by Lavoisier—as inconsistent with the belief that, when a calx turns into a metal, it is because phlogiston is escaping.

Any *novel* anomaly must weaken a theory, and hence make challenges to other aspects of the theory easier, and perhaps revive (by making it starker, or by otherwise bringing it freshly to attention) some other previously well-tamed anomaly. Hence, both directly and indirectly, new evidence or argument that is jarring (given the expectations generated by the prevailing view) can make a previously robustly entrenched habit of mind vulnerable. But if beyond a barrier there is only a further barrier—not a striking novelty that motivates the persistent effort required to break an entrenched habit—then the opportunity to go further will ordinarily be lost. The anomaly must eventually cease to be seen as jarring, and the barrier will be secure until some further anomaly opens another window of opportunity.

Even while an anomaly is effective, it will not prompt conscious effort to break a habit of mind, since ordinarily there is no explicit awareness that a habit of mind is involved. But sufficiently striking anomaly does prompt repeated effort to see the situation (or at a later stage, to recapture a momentary glimpse in which it was seen) in some way that makes sense, that is coherent, that somehow resolves the cognitive tension of an intuition that slips disconcertingly between looking right and looking wrong. A critical habit of mind (responding in the usual way) yields an intuition that looks right, but if some chance cue makes the anomaly particularly salient, the same intuition now looks like it might be wrong. If occasional breaks from the usual habit produce striking results, the effort to hold on to the novel view can be sustained. Otherwise, the anomalous arguments or evidence will be tamed. Having noticed something that does not make sense in terms of usual views, a person is sensitized to such effects, and more easily notices others. But sustaining that cumulative process depends on breakthrough effects, that is, on seeing glimpses of ideas and effects too striking to be quickly forgotten.

In Lakatosian language, but with a meaning different from that of Lakatos (since we are talking of tacit habits, not explicit rules and principles), the normal response to anomalies is some change in the protective belt, not a disruption of the hard core. Most anomalies— indeed, the vast majority—merely catch some oddity of a particular sort of circumstance. So it makes Darwinian sense that cognition would be such that, absent striking breakthrough effects, anomalies gradually lose their power to shock, and hence to incite consideration of radically un- conventional ideas. An anomaly once tamed might still be revived—by seeing it under more striking conditions, or in the context of some novel related anomaly, or by a person new to the field and hence not so thor- oughly habituated to it, or (most usual) by some combination of these. But without striking breakthrough effects, that will only be a temporary revival.

3.5

Now we come to the aspect of the argument that most readily strikes even sympathetic readers as implausible: I will try to show why we can expect that, characteristically, a *unique* barrier—some particu- lar habit of mind—is the key to the emergence of a revolutionary idea. This prompts objections along these lines:

Even if habits of mind in fact play the constitutive role urged here, why should it follow that some particular habit of mind is ordi- narily critical for a particular paradigm shift? It must be a network of interacting habits that is characteristic of a paradigm, and hence a net- work of habits that must change. Why not suppose that the empirical anomalies or whatever else serves to challenge what had been taken for granted might develop at a number of different points in the network, or in several more or less at once, with no particular habit of mind destined to play a uniquely critical role?

And even if we are only concerned with what happened in some particular historical case (not with abstract might-have-beens), why not even then suppose multiple points on which the discovery proceeded? Again, even if there was a single point of initial escape from the commitments of the prior paradigm, might that not be something merely idiosyncratic for the particular discoverer in particular circumstances, and hence something that, even if identifiable, might be of no real significance for understanding the development of science or for understanding more generally the emergence of radically novel ideas?

Summing up these objections using an analogy from *Patterns:* if discovery is like finding a path to the top of a mountain, why suppose that any feasible route would encounter a certain barrier, or that one particular point of difficulty on the path actually taken (even if that was somehow the only path that might have been taken) was uniquely *the* barrier that was critical?

As usual in such a situation, there are two sorts of answer: a conceptual argument for the plausibility of the claim, then an attempt to show successful application of the argument. The main business of the balance of the study is to develop applications. The conceptual argument, which is simple and in fact almost tautological, follows here, with the various possible cases illustrated (but not argued in detail here) by some aspect of the Copernican case.

Consider a case in which we can point to multiple barriers to some radical novelty. When multiple barriers are involved in the same phenomena, they will always be to some extent interdependent: if one were broken or even weakened, the other would become more vulnerable. Conversely, so long as both are in place, even the intrinsically less difficult of the barriers may be effectively unchallengeable. The nested-spheres and central-Earth habits of mind I will take up when we come to detailed treatment of the Copernican case provides a very well marked example. The heliocentric idea obviously cannot make sense until the powerful physical sense of the Earth as the fixed base from which all other motion is seen has been effectively challenged. But the nested-spheres habit of mind that is built into the practice of Ptolemaic astronomy blocks seeing the striking connections that come from allowing even the nonterrestrial planets to orbit the sun, as described in detail in *Patterns* [12.2]. Once on the scene, these lead to surprising anomalies for central-Earth intuitions. But so long as central-Earth intuitions are solidly in place, they block these insights. On the other hand, once someone had gotten past the nested-spheres barrier, the central-Earth barrier could also be challenged. Even for astronomers like Tycho who remained committed to the central Earth, it became something that was open to doubt and had to be defended by explicit arguments: it was no longer a point so obvious that even the possibility that it could be wrong would never be taken seriously.

So although it is possible and perhaps even common that there would be multiple barriers, pragmatically we might still be able to identify a unique barrier, with the property defined earlier: it is *unique* in the sense that a person competent to make the discovery who somehow gets beyond that barrier is likely to go on to make the discovery, and not so otherwise. And the structure of the issue, as in the Copernican case, may be such as to determine that breaking of one potential barrier could only plausibly come after a prior barrier had been challenged.

In general, suppose that there are indeed two candidates for barrier. We can still expect one particular habit of mind to be critical, as can be seen from considering the full range of possibilities, using some aspect of the Copernican case to illustrate each of the possibilities.

1. Suppose that anomalies arise (I mention again that new arguments rather than new evidence may be most crucial), sufficient to at least momentarily make some ordinarily solid intuition open to doubt, but there are no immediate breakthrough effects. So momentarily getting past one but not both of the barriers yields no striking further evidence or argument. Then the anomaly is almost certain to be merely tamed, and both barriers will remain in place. This must have happened repeatedly in the Copernican case, since Aristarchus's conjecture was never wholly forgotten, and over the centuries many people first encountering this startling idea must have been led to ponder the possibility that he could be right. But so long as the intuitions remained solidly in place, thinking about that possibility did not suggest striking consequences, but only that it could not possibly make sense. From all the centuries between Ptolemy and Copernicus, nothing indicates that anyone was able to take the possibility seriously enough to say something new on it.

2. Or suppose that there are breakthrough effects but a second barrier remains strong anyway. The breakthrough effects, although strong enough to lead to a definite break with the first barrier, are not of the right sort to provide a sufficient challenge to the second barrier. Then the breakthrough effects are remembered as an important discovery. And the second barrier would be more vulnerable. But now there is no contiguity in time linking the two into a single episode. When the second barrier finally falls, it will be as part of an eventual second, independent, episode. In Kuhnian language, for this possibility we would have one paradigm shift with the first breakthrough, and another later on with the second. A good example is the challenge to Ptolemaic cos-

mology from Copernicus, then decades later the abandonment of the Ptolemaic technical apparatus after Kepler's break with the commitment to UCM. There is a unique barrier for each episode.

3. There is only one remaining possibility, which also supposes that there are breakthrough effects that are now sufficiently discomforting for the second barrier that it has lost its potential to block a remarkable new idea. So long as both potential barriers are fully in place, they reinforce each other. A momentary break with one is particularly difficult to sustain, since the second is there to help assure that the break will be quickly seen as a mistake. But if the breakthrough effects are sufficient to lead to a sustained challenge to the habit of mind as an effective barrier, the second barrier no longer has this automatic support. And, as I have been stressing, the breakthrough effects from breaching the first barrier themselves may stand as anomalies challenging the second barrier. The joint effect is, for this case, that effectively there was a unique barrier: once it was broken the second barrier, though perhaps ex ante even stronger than the first, as we may suppose the central-Earth barrier was even more robust than the nested-spheres barrier, now is no longer an effective barrier. I have already said something about this Copernican example. Later on, I go through it in the detail it deserves, providing evidence beyond that of *Patterns* that indeed Copernican discovery started from a break with nested-spheres commitments.

Overall, then, for all three possibilities, although multiple habits of mind may be critical for some idea to emerge we still have, for each episode of radical discovery, one barrier.[6]

This dynamic process can be worked out in a good deal of detail, exploiting on the one hand the account of "rivalry" and "taming" developed in *Patterns* (chapter 7), and on the other hand the parallels between habits of mind and physical habits worked out in the opening chapter of this study. In the balance of the study, I try to make some useful progress on developing those parallels and their interactions, culminating in the discussion (beginning in chapter 10) of *economy* and *comfort* as properties favored by the endemic cognitive propensities that ultimately govern the formation and hence also the breaking of habits of mind.

Four

The Overthrow of Phlogiston: 1

For the Copernican discovery, the great puzzle is to see how anyone first managed to believe that the Earth is a planet flying through space. But for the overthrow of phlogiston, the puzzle is to understand why men as able as Priestley and Cavendish, and indeed Lavoisier himself, found it so hard to give up the idea of phlogiston.

The case is a salient one for the barrier analysis. The symptoms of a Kuhnian revolution are well marked. Both Lavoisier and his adversaries explicitly saw the overthrow of phlogiston as revolutionary. But a feature of recent writing on the episode has been skepticism that giving up phlogiston, of itself, could be taken to be revolutionary.[1] We have a sharp disparity between the evident cognitive difficulty of the change when new versus the logical triviality that is seen now. Further, since we are all familiar with seeing flames leap up and dissolve into the air as the burning fuel gradually disappears, there is a salient suggestion about the basis of a habit of mind that (on the barrier view) might account for the disparity between cognitive and logical difficulty.

The barrier analysis turns out to yield an account that departs substantially from what can be regarded as the "usual story." We also get a sharp illustration of a useful distinction between a "normal" and a "Kuhnian" revolution, where the former is revolutionary from the gap viewpoint but gives no marked sign of Kuhnian incommensurability. The latter may not seem logically difficult, but is defined by plentiful evidence of incommensurability.

4.2

The "usual story" starts in 1772 with Lavoisier's discovery that combustion and calcination are processes in which large quantities of air are absorbed, explaining why the products of these processes weigh more than the original substance. By 1776 Lavoisier and others had identified a particular fraction of ordinary air responsible for the increase in weight: what Priestley called "dephlogisticated air," Scheele called "fire air," and Lavoisier eventually called "oxygen."[2]

Of the three discoverers, only Lavoisier eventually saw that all the processes that had previously been explained in terms of a loss of phlogiston might now be explained by a gain of oxygen. He showed that calcination in pure oxygen continued until all the gas had disappeared; or if there was sufficient oxygen that the supply of metal available to be calcined was what limited the process, then the remaining gas turned out to be only pure oxygen.[3] Either way, the weight gained by the calx equaled the weight of the gas that disappeared.

For the special case of mercury, we get what in the usual story is a decisive demonstration of the fallacy of the phlogistic view. Mercury can be calcinated by heating, absorbing oxygen, gaining weight equal to the weight of the oxygen that disappears. Nothing detectable is given off in this process. And by more intense heating the process can be reversed, yielding back flowing mercury and the original weight of oxygen. So it is oxygen added, not phlogiston driven off, that turns the metal into its calx. And it is oxygen driven out of the calx, not phlogiston added, that yields metal from the calx.

Lavoisier reported compatible results from experiments on combustion, fermentation, and respiration. Visible fire, he argued, came from heat released when oxygen combined with the combustible material—a process that was demonstrably occurring, so that there was no need to hypothesize an additional process (escape of phlogiston) to produce the heat and flame of combustion. In short, the simplest explanations of phlogistic processes involved no reference to phlogiston. So what reason was there to suppose that phlogiston existed?

On the usual account, Lavoisier's 1777 argument against phlogiston was straightforward. Conant (1950) gives the dates of the chemical revolution as 1772–78, and nearly all accounts are similar, treating later evidence as of secondary importance and most often ignoring it entirely. But a few writers who have given detailed attention to the response to Lavoisier after 1777, notably Toulmin (1957), Musgrave (1976), and Perrin (1988), find that things were not so straightforward. Perrin shows that serious controversy did not begin until 1784, seven years after Lavoisier first made a case against phlogiston. His argument then turned substantially on new evidence. Toulmin and Musgrave, against the grain of most accounts, argue that in fact there were perfectly good pragmatic grounds for resisting Lavoisier's claims in 1777, and indeed argue (much less plausibly) that such grounds existed even after 1783.[4]

On the usual view, an absence of any further evidence or argument from Lavoisier for six years after 1777 goes unnoticed, as if nothing more was needed. And indeed, if Lavoisier's 1777 argument was obviously right, why would he need to repeat himself?[5] Conant (1950, 53) interpreted the renewed effort by Lavoisier after the water discovery as due to a *revival* of the phlogistonist case—the opposite of what I will be arguing here.

4.3

Lavoisier has often been interpreted as discovering what phlogiston really was, namely, *negative* oxygen. So of course Lavoisier's argument worked better than the phlogistic alternative, since it necessarily explained everything that invocation of phlogiston explained, and also explained why substances gained weight when they lost phlogiston, why the weight gained just equaled the weight lost by the air, why the air released when the process was reversed supported combustion far more energetically than the original atmospheric air.[6] But although all of this was in hand for Lavoisier by the end of 1775, his doubts about phlogiston did not emerge until his first public attack on phlogiston in late fall of 1777 (Holmes 1985, 30–56). Even then, Lavoisier was tentative in his presentation, allowing that his theory was not completely worked out, and that it was presented only as a hypothesis.[7] He promised a series of papers developing his case. But it was seven years before any prominent chemist supported Lavoisier, or before Lavoisier himself seriously took up the extended discussion of the issue he had promised.

All this is particularly curious since the argument Lavoisier made does not seem to have lacked force for readers interested in science but not deeply involved in the practice of chemistry. Perrin (1970) describes two articles exploiting Lavoisier's own evidence to attack phlogiston several years before Lavoisier himself was moved to do so. Priestley had to respond to skepticism he found among his nonchemist Lunar Society friends in Manchester.

We need some explanation, therefore, of the critical point that is muted by the usual account, which is that even Lavoisier himself was slow to make arguments against phlogiston, and that when he did give an argument that seems convincing today, no chemist followed. Rather, Lavoisier seems to have taken his flag down and turned his attention to other matters for the better part of the next decade.

4.4

The papers of Priestley, Scheele, and other important defenders of phlogiston show that Lavoisier's basic explanation of why calxes gained weight was never a matter of controversy. By the mid-1770s no leading chemist supposed that phlogiston had negative weight or that calxes gained weight by absorbing particles of fire, or confused the decrease in density of calxes with the increase in weight. As Perrin (1987) observes, the period from Lavoisier's initial contribution in 1772 to his argument against phlogiston in late 1777 was one of accord, not controversy.

A variety of detailed interpretations were offered of just what was happening when a calx absorbed air, but that the absorption of air accounted for the weight gain was not at issue, so that Lavoisier could not gain support merely by pointing out that his discovery resolved that old puzzle. I have already mentioned that it was not enough to move even Lavoisier himself to attack phlogiston until five years after that discovery was in hand, nor for two years after he (and Priestley and Scheele) knew that it was a particular component of ordinary air that was absorbed.

After the cool reception of his 1777 presentation, Lavoisier was silent about phlogiston until 1783. Notes apparently dating from this period outline the topics apparently intended to fulfill his promise of a set of papers giving the details of his theory. But this program was laid aside, and indeed for about four years after announcing his new theory, Lavoisier seems to have largely given up work on chemistry (Holmes 1985, 129–30).

But in June 1783 Lavoisier learned of Cavendish's still unpublished discovery that inflammable air (hydrogen) exploded with dephlogisticated air (oxygen) yielded pure water. He *then* renewed his attack on phlogiston, with an aggressiveness far beyond his position in 1777. Within a couple of years he was speaking without qualification of phlogiston as a useless idea that had to be cleared away if chemistry was to make progress.[8]

After hearing Lavoisier's initial (1777) attack, Macquer (the grand old man of French chemistry) wrote to reassure de Morveau (eventually a key ally of Lavoisier, but not until 1785) that Lavoisier's argument did not amount to much (Holmes 1985, 127). The first major publication that treated Lavoisier's claims as a serious threat is Kirwan's ([1784] 1789), which begins with a historical introduction that illustrates very well the situation I have been trying to describe. Until the surprising discovery of the composition of water, Kirwan writes, the situation was admirably clear. He commends Lavoisier for his discovery of why calxes gain weight. Kirwan treats that as a definitive resolution of a long-standing anomaly, but one that in no way threatened the basic phlogistonist story. Further (Kirwan says), understanding of phlogiston had advanced considerably beyond Stahl's original important insight into the relation between combustion and calcination. Now this previously elusive substance can be isolated, weighed, and experimented with, just like other substances. We will see in a moment why Kirwan made that claim.

Perrin (1988) compiled a diagram of how events unfolded, for purposes unrelated to the argument here. But it serves very well to illustrate the basic points being urged. Perrin's periodization of the story is exactly that suggested here, and his detailed survey of French chemists shows strikingly the coincidence between the period of open conflict that begins with Lavoisier's paper on the composition of water read to the Academy late in 1783, and the beginning in 1784 of what soon would become a flood of converts to his view.[9]

4.5

Stahl's original demonstrations provided striking evidence that the fire that could be seen escaping during combustion, for which he used the label *phlogiston,* could be manipulated by chemists, transferring it from one material to another and thereby transforming the substances involved. What had until then been considered unrelated processes (in particular calcination and combustion) could now be seen as different forms of the very same process. Calcination was seen, for the first time, to be simply slow combustion. Stahl could very strikingly illustrate the process he proposed. He would heat substances and apply a bit of charcoal, which could be seen to be very rich in phlogiston since, once combustion was complete, only a bit of ash was left. A person could see with his own eyes that, as phlogiston was absorbed at the spot where the charcoal had been applied, sulfur emerged from sulfuric acid, lead from calx of lead, and so on. Lavoisier could say the same thing, except that that he (as we today) would say that oxygen was absorbed from the calx, not phlogiston absorbed into the calx.

Lavoisier challenged Stahl's interpretation in 1777 and with evidence that looks convincing to us. But his argument, as I have stressed, was tentative and had little effect. What, then, was missing from the case against phlogiston prior to the mid-1780s, and how did the discovery of the composition of water remedy the situation?

A fundamental point is that the overthrow of phlogiston took place in the context of a more general story of the development of "pneumatic chemistry," which turned on the recognition that "airs" were essential for understanding how material substances are compounded, are separated, respond to heat, and so on. Chemistry, we would say, finally came to include gases as well as solids and liquids.[10]

Although there is a significant prehistory, a fair date for when pneumatic chemistry emerged as a major component of chemical research would be 1766, when Cavendish's papers on what he called "factitious airs" (gases extracted from solids and liquids) were read to the Royal Society, winning Cavendish the society's gold medal. This emergence of pneumatic chemistry required far more of a *conceptual* revolution than the overthrow of phlogiston, which logically was only an episode in the larger story. Just that perception is reflected in the recent surge of articles questioning that the overthrow of phlogiston could be as revolutionary as Lavoisier and his adversaries supposed, and as the earlier history of science had allowed.

For the transition to pneumatic chemistry to occur, a familiar *substance* (air) had to be seen as a novel *category*. Heat any solid sufficiently, on the new view, and it will eventually become an air. Many new substances appear—not discovered in out-of-the-way places, but seen to have been always around, fixed in everyday substances, and released by familiar processes. New technologies, new skills, and new work habits now played a leading role in chemistry, as investigators sought to manipulate, weigh, store, identify, and characterize this proliferation of novelties essential for understanding chemical processes. A most important by-product of this evolution was that the balance could at last play a critical role in chemistry. Until this pneumatic revolution, mass was not conserved in chemical processes as they were studied (that is, ignoring the role of gases); hence the balance could not be a critical instrument.[11]

The overthrow of phlogiston would seem to be an easy and congenial component of all this. As has often been pointed out, Lavoisier's new idea is effectively that phlogiston turns out to be negative oxygen. Even his most conspicuous error is a direct reflection of that. On Stahl's theory, acids are elementary, so that heating an acid with charcoal (almost pure phlogiston, since almost nothing is left after its matter of fire has escaped) yields sulfur, which is a compound of sulfuric acid and phlogiston. Lavoisier simply turns things around: sulfur is elementary, and adding oxygen is essential to compounding it into its acid.

In general, everything learned about phlogistic processes holds, except that chemists now say that oxygen is gained instead of phlogiston lost, or the converse. In place of an elusive substance with puzzling properties, we have a well-defined entity that can be extracted, examined, manipulated, weighed, even *breathed!* (Priestley suggested at the end of his 1775 paper on the discovery, that inhaling the new gas might become "a fashionable article in luxury. Hitherto only two mice and myself have had the privilege.")

Recognizing oxygen as what really lay behind the ability of phlogiston to lend coherence to a variety of superficially unrelated phenomena (respiration, fermentation, combustion, calcination) also simultaneously—that is, at once, with no further information—resolved the puzzles of why calxes gain weight and why phlogiston had been so hard to isolate (chemists had been looking in the wrong place), and of course it explained why the calx of mercury could be reduced without access to charcoal or any other observable source of phlogiston. It was, in short, the sort of new idea that we could expect to have a great deal of appeal. Logically, Lavoisier's proposal should have been quite capable of competing with the reinterpretation of phlogiston emphasized in the passage quoted from Kirwan [4.4], and to which we need to return.

But though the transition from talking about phlogiston to talking about oxygen *logically* looks to be exhilarating, not difficult, *cognitively* it was plainly painful for most chemists. And looking now to the core idea of the barrier analysis, we can see why that could be so. If what is guiding intuition is at bottom the physical sense of something escaping as flame leaps from a fire and the fuel gradually disappears, then however neatly oxygen fits the logic of the situation, Lavoisier's story violates that (tacit) physical sense of what is going on.[12]

Priestley (1796) began his last major attempt to make his case for phlogiston with the remark that "there have been few, if any, revolutions so great, so sudden, and so general, as the prevalence of . . . the new system of chemistry . . . over the doctrine of Stahl . . . [which] has been constantly and uniformly advancing in reputation more than 10 years." But later in the same passage comes the remark that "every year of the last twenty or thirty has been of more importance to science and especially to chemistry, than any ten in the preceding century."

So we have unambiguous references to two overlapping but distinct periods. One is the period of the overthrow of phlogiston ("more than 10 years"); that is, since the mid-1780s, or as I will be stressing, since the discovery of the composition of water. The other period Priestley mentions is the longer span that covers the rise of pneumatic chemistry, with its attendant radical innovations in the way the work of chemists was done and in the way chemical processes were being understood.

In the language I have been using, we have a Kuhnian revolution: cognitively difficult though logically not so, hence best understood as turning on the presence of a barrier habit of mind. And that takes place within the context of a fundamental reconception of chemical processes to take account of the role of gases. That latter transformation of concepts is logically far more comprehensive and radical, but cognitively easier, in just the way discussed in the theoretical argument of chapter 3. However imposing the *gap* between it and previous ideas, no effective *barrier* habit of mind happens to be encountered in the transition.

4.6

The rise of pneumatic chemistry inevitably yielded a readiness to accept new understandings of old puzzles. The world had changed for chemists in a drastic way. No one is surprised to find that drastic change has important consequences. I have already repeatedly referred to the ready acceptance of Lavoisier's solution of the puzzle of weight gains (understood, of course, as an independent discovery, with no immediate connection to the essential nature of phlogistic processes). It is scarcely surprising then to notice that, however uncomfortable it might be to give up phlogiston entirely, the chemists of the period were not at all uncomfortable with new ideas about the nature of phlogiston.

Pneumatic chemistry indeed had provided new ideas about phlogiston even before Lavoisier or Priestley began the work that would make them the leading chemists of the time. Cavendish's prizewinning 1766 paper was already linking inflammable air from metals with the long-elusive phlogiston. Although his position later became more complicated, by the time Lavoisier offered his tentative first attack on phlogiston (1777), it simply was no longer the case that a demonstrable substance—you could weigh it, experiment with it, breathe it—was being offered to fill the role hitherto attributed to a ghostly substance with none of those possibilities. Aside from breathing, phlogiston might have all those properties (as Kirwan emphasized), since even chemists skeptical of Cavendish's identification of phlogiston with inflammable air (hydrogen) had that possibility before them.

This new interpretation of phlogiston was based on pneumatic experiments predating by a decade Lavoisier's first public doubts about the existence of phlogiston. Hence a particular weakness of Lavoisier's case was its failure to explain Cavendish's apparent demonstration that phlogiston in fact existed: you could bottle it. What Cavendish had seen as phlogiston (or, later, something more complex, but with phlogiston the central ingredient) was produced in a way that could be read as supporting Stahl's account in a striking way. Dissolving various metals in a weak acid, Cavendish reported, yields a gas as the metal disappears. The residue when the acid is evaporated is a salt indistinguishable from the salt obtained when the calx of the metal is dissolved. If subjected to the treatment that would recover the calx when calx is dissolved, this salt from the metal also yields what appears to be the calx. So on the phlogistic argument we would expect that the dissolving metal had lost its phlogiston, yielding the calx, which continued to react with the acid, yielding the salt. Hence the theory implied that, when a calx obtained by this route was retrieved, it should be possible to restore it to its original metallic form by heating it with charcoal: this would allow the calx to combine with phlogiston, as the same technique in general produced metals from their calxes, produced sulfur from sulfuric acid, phosphorus from phosphoric acid, and so on.

If reviving the metal in this way worked—as indeed it proved to work—then the gas that had escaped should be the elusive phlogiston, finally revealed by the novel techniques of pneumatic chemistry. If so, it should burn, as indeed it did (hence the name, "inflammable air").

Priestley eventually believed he completed the demonstration by showing that the process of extracting inflammable air from a metal could be reversed. He heated a calx in a bell jar filled with inflammable air. Parallel to Stahl's demonstration of the appearance of drops of liquid metal where he applied bits of charcoal to a hot calx, the inflammable air in Priestley's experiment gradually disappeared (apparently absorbed by the calx), and the calx was converted back to metal. However, although everything here seemed to go exactly as a defender of phlogiston would hope, this experiment was not performed until 1782, so that it cannot account for the failure of Lavoisier's 1777 claims. Indeed, as will be developed in chapter 5, what Priestley belatedly presented as an absolute proof of the existence of phlogiston quickly turned into a contributor to its absolute downfall.

Another phlogistic argument was less striking but more timely. Under some conditions an inflammable air could be extracted from charcoal, which made sense for a phlogistonist, since charcoal was presumed to be almost pure phlogiston (leaving only a scant ash after combustion was complete). But the inflammable air from charcoal was much heavier than inflammable air from metals, burned gently rather than explosively, and yielded fixed air (carbon dioxide), while the product of burning inflammable air from metals was elusive. So an opponent of phlogiston could doubt that this heavy inflammable air was in fact the same substance as the light inflammable air from metals; or he could allow that heavy inflammable air might indeed be a variant of (or compound of) the same substance as light inflammable air, but argue that that possibility did not imply that every calcination and combustion involved the escape of some common principle of combustion. Lavoisier himself seems to have wavered between the two. This issue proved difficult to clarify, and it was after the turn of the century before "heavy" inflammable air (carbon monoxide) was clearly distinguished from ordinary inflammable air (hydrogen). The fate of phlogiston had by then been effectively settled for at least a decade.

Since the contest was not really joined until 1783, we have little evidence of how defenders of phlogiston would have responded had Lavoisier pushed his position. But overall, Cavendish's unanswered claim that phlogiston could be demonstrably extracted (as inflammable air) from metals, plus a variety of less striking or more contested evidence (such as the extraction of inflammable air from charcoal just mentioned), allowed defenders of phlogiston substantial grounds for doubting that the discovery of oxygen implied that phlogiston was redundant.

The experiment with calx of mercury, which is ordinarily treated as decisive, was explained by Priestley as showing merely that a particular form of one calx had the odd property that the phlogiston would stay with the calx instead of escaping. Contrary to the usual story, which treats the absence of explicit debate as showing the defenders had no answer to Lavoisier, the actual situation was that Lavoisier apparently lacked sufficient confidence in his own argument to make a determined effort. But allowing that there were what even today can be regarded as good grounds for resisting Lavoisier's claim in the years immediately following 1777 does not mean that there is no substantial puzzle. On the contrary, there was a very striking puzzle.

4.7

If science proceeded in a narrowly rational way, Lavoisier's initiative would have prompted intense discussion. In particular, it was known that sulfur, phosphorus, and even some metals could be burned in a sealed container filled with oxygen, after which any remaining gas was still only pure oxygen. If the visible flame in these processes was phlogiston escaping, it had somehow disappeared. The weight of the container was unchanged. This encouraged the idea that phlogiston was escaping as heat or light, which had no weight. Yet the phlogiston apparently extracted from metals (the substance Lavoisier eventually named hydrogen), though very light, had positive weight. Candles or lumps of charcoal, which had been taken to be made mostly of phlogiston, since almost nothing remained after combustion, certainly lost weight as their bulk diminished. So apparently phlogiston both had weight and did not have weight. It was these contradictions that Lavoisier would dramatize when he returned to his attack on phlogiston after 1783, leading to the often-quoted passage calling phlogiston a "veritable Proteus, changing form from moment to moment." But before 1783, Lavoisier used no such drastic language, and chemists gave no sign that they perceived any crisis.

Lavoisier himself tried to find the product of burning inflammable air—which would greatly strengthen his case, since whatever it was couldn't be found after the experiments with combustion in pure oxygen. But Lavoisier could find no such product, a disappointment that left him in no position to attack the phlogistonists for failing to show any detectable effect of the phlogiston escaping. Presumably this contributed to his failure to follow up the initial attack on phlogiston after 1777. Certainly Lavoisier's failure to effectively pursue this lead suggests that he had lost some confidence in his argument. Giving up an investigation when difficulties arose, rather than aggressively pursuing things, was very much out of character for him.

Lavoisier's performance elsewhere shows that he was perfectly familiar with how to pursue this matter: inflammable air (the purported phlogiston) should be ignited within a closed container, and the container weighed. If the weight was unchanged, then a careful search for the combustion product was mandated. For no one ever supposed that weight could exist independent of some substance. In fact we know that weight would not be lost and that the search for a combustion product (water!), once vigorously pursued, would have been a short one. We will see that Priestley, pursuing a similar investigation but hoping to find that weight was lost, let himself be misled into thinking he had found what he was hoping to find. But Cavendish, though a phlogistonist, did not make that error.[13] And it seems unlikely that a Lavoisier with confidence in his own theory would blunder into an error that contradicted his own convictions. On a rational reconstruction the discovery of the composition of water should have come promptly, given the conflicting empirical results in hand by 1777. But it did not happen that way.

4.8

Until the 1790s, when he found himself almost the last holdout against Lavoisier, Priestley was almost continually reminding his readers of how surprised he was by his results. That is a rhetorical mode that is not used even by someone as candid as Priestley in a context where the conservative response "so it's probably a mistake" is a serious risk. Rather, all indications are that we are dealing with a community expecting surprising new results. But, on the record, until late 1783, no leading chemist was prepared to follow Lavoisier in giving up phlogiston, and Lavoisier himself was not pressing the point.

If I toss a coin and it comes up heads, I need no special explanation. Even several heads in a row do not demand an explanation. But a long string of tosses, all of which turn up heads (here, many chemists' judgments all favoring phlogiston though logically the situation was ambiguous), suggests that something more than the superficial logic of the situation is needed to account for what we see. On the barrier view, the nonrandom factor in the situation was the powerful disposition to see as somehow wrong a story that fails to include a component satisfying the deeply entrenched, habitual sense that something is escaping from a burning substance, which had been further deeply entangled in the practice of chemistry through Stahl's brilliant extension of phlogistic ideas and demonstrations to calcination and other phenomena that indeed we still see as intimately related to combustion. Given that deeply entrenched and entangled phlogistic propensity, it was hard to see as anything but obviously right a piece of evidence (from the behavior of metals in acids diluted by water) that seemed to confirm that intuition. The peculiar fact that inflammable air was obtained from metals only when dissolved in *dilute* acids did not provoke aggressive follow-up, even from Lavoisier though his own theory logically suggested experiments that would yield striking results.

So the water discovery came later, and serendipitously from a convinced phlogistonist (Cavendish) who continued to defend the old view, shaping his account of the discovery to fit. Lavoisier eventually performed the critical experiments, not because he was moved by his own 1777 argument, but only when Cavendish's discovery fell into his lap. Unless you are able to believe that if only Lavoisier were as smart and brave as you or I he would not have had this problem, there is on the face of things a considerable argument here for the importance of entrenched barriers.

Five

The Overthrow of Phlogiston: 2

Around 1780 Priestley and his friend Warltire believed they had shown that heat had a detectable weight, which, if true, would be a comfortable finding for defenders of phlogiston, providing a resolution of the conflicting evidence on this point [4.7]. Priestley and Warltire weighed a jar containing a mixture of inflammable air (hydrogen) and dephlogisticated air (oxygen), exploded the mixture with an electric spark, let the bottle cool, and reweighed the jar, finding (they believed) a small weight loss. They also noticed the appearance of a certain amount of dew inside the bottle. But they attached no importance to it. Priestley and other chemists had noticed that before. It would be fairly common, since (we know today) essentially all organic materials contain hydrogen and so must produce water on combustion. But in the 1770s, the anomaly could be tamed by the conjecture that airs in general invariably contained a certain amount of water as an intrinsic component of the gas, aside from any humidity that could be eliminated by preheating the material or passing the gas through moisture-absorbing crystals.

Cavendish undertook to replicate the Priestley and Warltire experiment on a larger scale. He could not find the loss of weight they reported. But he became interested in the dew that appeared, and embarked on a careful investigation that showed that an appropriate mixture of inflammable air (or phlogiston as Cavendish also sometimes called it, or hydrogen as Lavoisier eventually called it), plus dephlogisticated air (oxygen) would explode, leaving no gas at all, but only pure water just equal to the combined weight of the two original gases.[1]

Priestley and others in England knew of this work by early 1783. But Cavendish still had not published anything when his assistant Blagden gave the news to Lavoisier in June 1783. According to Blagden, Lavoisier was startled and skeptical. Lavoisier (Blagden reported) thought that burning inflammable air should produce fixed air ("aerial acid").[2] For Lavoisier—and many of his adversaries as well, including both co-discoverers of oxygen, Priestley and Scheele—expected that oxidation always yields an acid.

Lavoisier's expectation here would be encouraged by the coincidence that heavy inflammable air (carbon monoxide)—or even light inflammable air contaminated with carbon monoxide—did produce aerial acid (carbon dioxide), and mixtures of light inflammable air (hydrogen) and oxygen with an excess of oxygen produced a dilute nitric acid. This came from the inevitable contamination of the gases with the nitrogen that makes up 80% of common air. The failure of Lavoisier's experiments with burning inflammable air was linked to his expectation that the combustion should yield an acid. His setup, focused on identifying the acid he expected, involved water in the combustion chamber (enough to swamp the water produced by the combustion), so that Lavoisier missed Cavendish's discovery.[3] Yet giving up so easily was not Lavoisier's usual way. When he was confident he was on the right track, he was ingenious and persistent. As I have already stressed [4.7], passive response to failure here raises strong doubts about Lavoisier's confidence in his own argument against phlogiston.

The news from Blagden radically changed the situation. However skeptical his first reaction, Lavoisier quickly realized the significance of the possibility Cavendish was right. Before the week was out he had crudely replicated Cavendish's result (with Blagden and others as witnesses) and had announced the discovery of the composition of water to the Academy, neglecting to mention Cavendish. He went on to replicate Cavendish's work with more care and added to it a demonstration that the process could be reversed: he showed how to extract hydrogen from water, while the iron used in the process became oxidized, with the gain in weight of the iron plus the weight of the evolved hydrogen equaling the weight of the water that disappeared. All this was reported in detail to the Academy the following November, this time with some credit to Cavendish.

Quite suddenly, then, Lavoisier's worst problems with his attack on phlogiston were resolved. He now had strong grounds to suppose that the inflammable air that appeared when a metal was dissolved came from decomposition of water, not from the metal, and left the defenders the puzzle of how to make a plausible case for the contrary view. For Lavoisier now could explain why the gas evolved from dilute solutions of the acid but not from strong solutions.[4] And when he was informed by Blagden of Priestley's demonstration that a calx could be reduced by apparently absorbing inflammable air—what Priestley regarded as a decisive demonstration favoring phlogiston—it was obvious that Priestley's marvelous demonstration should be repeated over mercury (instead of over water), which would show that water was being produced as the calx revived (the hydrogen was combining with the oxygen from the calx to form water, not with the calx to form metal). If the calx were revived with charcoal, no water was produced, but rather fixed air (carbon dioxide), as Lavoisier's theory required.

Priestley had believed his experiment proved "beyond all doubt" that phlogiston was real and a component of metals.[5] But quite suddenly the same experiment yielded a new interpretation with the opposite consequences. Once the composition of water was known, the result exactly fitted the claims Lavoisier had put forth six years earlier, and it was the defenders of phlogiston who were now confronted with a demonstration not easy to answer and too striking to ignore. A substantial part of Lavoisier's paper reporting his analysis of water was devoted to a discussion of Priestley's experiment and to the radically contrasting interpretation he could give of it.

After 1783 Lavoisier increasingly moved beyond the tentativeness of his 1777 presentation, climaxing with the very aggressive tone of the "Reflections on Phlogiston" read to the Academy in the summer of 1785.[6] Starting in 1784 chemists began publicly to move to his side, and over the next few years Lavoisier's view grew markedly stronger and was soon dominant.[7] In sum, although nothing at all had come of Lavoisier's initiative for six years after his 1777 venture, things began to move rapidly after news of Cavendish's discovery reached Lavoisier at the end of June 1783.

In the most detailed response to Lavoisier, Kirwan remarked that the critical issue *after* Cavendish's discovery was whether in fact calcinable and combustible substances invariably contained inflammable air.[8] If so, then apparently the basic Stahlian insight was correct. As Kirwan put it, "The controversy is at present [circa 1785] confined to a few points, namely whether the inflammable principle [phlogiston] be found in . . . fixed air, sulphur, phosphorus, and metals" (1789, 6).

Now I have stressed that, in terms of the barrier view, accepting oxygen but nevertheless insisting that phlogiston also exists is cognitively understandable. If the fundamental habit of mind that is governing phlogistic intuitions is the salient one identified at the start of chapter 4, then Lavoisier's resolution of why combustion and calcination cause materials to gain weight doesn't directly challenge the sense that the flames emerging from burning material reveal something escaping. Further, as stressed in the general discussion [1.6], such a habit of mind will not disappear *merely* because its logical underpinnings have come into doubt.

By the mid-1780s (but not in his 1777 paper), Lavoisier was giving good evidence that he realized that it is the physical intuition of something escaping that had to be challenged. His remarks are now explicit and emphatic. Responding to Kirwan's remark just quoted, Lavoisier shows that he had a clear sense that it is futile to try to prove a negative (that metals and so on do *not* contain a common inflammable principle); but he could try to undermine the basis of the appeal of the positive belief.

> How is it that Stahl was led to suppose that an inflammable principle existed in combustible bodies? It is because he apprehended that there was no other way of explaining the disengagement of heat and light which takes place at the moment of combustion. It is to the separation or emission of the inflammable principle which was imprisoned in those bodies, but set at liberty in this process, that he attributes the effect. But in the hypothesis of Mr. Kirwan, [in many cases] there is no disengagement of the inflammable principle. . . . The phenomena of combustion cannot therefore be attributed to it, so that they remain without explanation. . . . What then does [Kirwan] gain by his supposition? Additional difficulties, and with no assistance to remove them."[9] (Lavoisier's comment in Kirwan 1789, 74)

5.2

Going at least back to Stahl, the best case for chemical claims was made by showing both analysis and synthesis. The belief that he had obtained such a two-way demonstration of the existence of phlogiston was what so pleased Priestley about his reduction of a calx with inflammable air.

Lavoisier was a keen pursuer of this tactic. Given that propensity, his own agenda, and Cavendish's result, the obvious next step (having produced water from hydrogen and oxygen) was to see if the process could be reversed, with oxygen and hydrogen produced from water. Here a salient indication of how to proceed was available in the commonplace knowledge that iron rusts readily when wet and doesn't when dry.

On Lavoisier's theory and Cavendish's result, when iron immersed in water rusts (absorbing oxygen), hydrogen should appear. Lavoisier started with simply putting iron filings in water and showing that indeed a gas was released as the iron rusted, and the gas was inflammable air (Lavoisier's hydrogen, his adversaries' phlogiston). So the simplest and most obvious experiment provided immediately striking results.

The matter soon became a spectacular instance of public knowledge, since about this time there was great excitement about the first demonstrations of manned flight, using hot-air balloons. An appealing alternative was to use hydrogen, which is much lighter than hot air. Lavoisier arranged to be commissioned to work out a way to produce hydrogen on the large scale that would be required, which he was able to do by passing steam through a red-hot cannon barrel, yielding rusty iron and a large volume of hydrogen (Holmes 1985, 211–12). The weight of water consumed equaled the weight of hydrogen produced plus the increase in weight of the iron. And, as Lavoisier stressed, the process was reversible. The inflammable air released would combine with a weight of oxygen equal to the weight increase of the calx to yield pure water with no residue.

How did Lavoisier's adversaries respond? Would it be plausible to see those responses as revealing a reasonable difference of judgment? Or do they reveal judgments that are somehow bizarre, other than in the light of something like the barrier argument?

Any presentation that is not extremely long must be simplified. In particular, I have left out the complications introduced by conflicting experimental results (chapter 4, note 3). What follows focuses the discussion on the response to Lavoisier's reinterpretation of the phlogistic demonstrations that metals contain inflammable air. Since these demonstrations were the showpiece of the phlogistic case, the discussion covers what seems to be the most important material.[10]

5.3

So long as the existence of phlogiston was an irresistible intuition for a person, there was a salient interpretation of the inflammable air experiments, and all the defenders provided some variant of it. A calx, as it appears to the eye, it was now said, was not a pure substance after all; it contained *water*. When a metal absorbed oxygen, the oxygen attracted the phlogiston that had been combined with calx to form the metal. This yielded water, but this water was absorbed by the calx.[11] On this view, Kirwan and other defenders argued that Lavoisier did not have the two-way demonstration he claimed to have. Where Lavoisier saw hydrogen attracting oxygen from the calx, forming water and metal, his adversaries saw phlogiston absorbed by the calx, forming metal and releasing water. Where Lavoisier saw oxygen attracted from water by a metal (forming a calx and setting hydrogen free), his adversaries saw water absorbed as the metal released hydrogen (phlogiston), leaving a calx and its associated invisible water.

The story can be made to work, but only in the sense of not being subject to some logical contradiction. But it is a clumsy story, which relies repeatedly on complicated conjectures to avoid what Lavoisier claimed—and indeed Kirwan explicitly allowed—were simpler explanations (Kirwan [1784] 1789, 10). Another leading defender (Cavendish) remarks that,

> As adding dephlogisticated air [oxygen] to a body comes to the same thing as depriving it of phlogiston [hydrogen] and adding water to it, and as there are, perhaps, no bodies entirely destitute of water, and as I know no way by which phlogiston can be transferred from one body to another, without leaving it uncertain whether water is not at the same time transferred, it will be very difficult to determine by experiment which of these opinions is the truest. (Cavendish 1784, quoted in Conant 1950, 54)

But by this time there was no evidence left that phlogiston existed (it had to be conjectured), and it was not enough for the phlogistic theory that calxes and so on might not be "entirely destitute" of water (itself a conjecture): a substantial fraction of their weight had to be hidden water. A calx became something that had never actually been seen, since in the revised phlogistic claim, the calx is the metal deprived of its phlogiston, but when this happens, it inevitably absorbs water. The phlogiston does not actually leave: it is still there with the calx, but combines with oxygen (dephlogisticated air) instead of with the calx. Or if it escapes, then water is absorbed from somewhere else. When fine-spun iron (like present-day steel wool) is burned in oxygen, flames seem to leap out of the material. But the phlogiston either remains behind, forming water with the oxygen that is being absorbed (leaving the flames unexplained, as Lavoisier complained [5.1]), or if the flames reveal phlogiston escaping, then water is merging with the calx from elsewhere. Among defenders of phlogiston all the possibilities had proponents.

Cavendish says it would be difficult for an experiment to decide whether Lavoisier was right, but that seems plausible only if what is required is something like a formal proof. It is always possible to "quine." But as will be discussed in a more general context (in chapter 10), we ordinarily find it easier to believe the more economical story, so the puzzle arises of why, for a few years at least, a number of very able and well-informed scientists (such as Cavendish) could find the conspicuously more complicated story easier to believe. The defenders' judgment was different from the normal propensity to believe the more economical story, and different from what these individuals (with the exception of Priestley) themselves saw within a few years. The issue for us is to account for the difference from usual behavior.

Of course the answer I want to suggest is that, for someone bound by the barrier habit of mind, phlogiston would not be *seen* as conjectural even though (after 1783) there was nothing that was even claimed to be direct evidence of its existence.[12] For someone with an entrenched intuitive sense that something was escaping when a substance burned (*phlogiston* being the label for that something), the prima facie simplicity of Lavoisier's scheme would hardly be appealing; for someone operating from the phlogistic side of the barrier, that was a *wrong* way to see things, all the more distasteful because of its apparent appeal. Rather, the person's intuition would be that, with work and insight, an account would eventually emerge in which this delusory simplicity of Lavoisier's account would dissolve. "We must not be deluded by a false show of simplicity," Kirwan remarks, "[since] when all is considered, the ancient doctrine will be found the more uniform of the two" (Kirwan [1784] 1789, 8).

5.4

By this point the phlogistonists had to defend a claim that one of the two constituents of water (phlogiston + oxygen) is the very stuff that purportedly is released when a metal changes to calx. What the metal purportedly absorbs (water) happens to contain as one of its constituents just what the metal is postulated to release (phlogiston), as if the magician was seen putting the rabbit in the hat, but then claimed the rabbit he pulled out was a *different* rabbit.

In terms of the phlogistic claims, there is no reason ex ante to expect that the weight of phlogiston (hydrogen) released from the metal should just equal the weight of phlogiston in the water that is simultaneously absorbed by the metal. It could have turned out to be half as much, or twice as much, or any other multiplier. But while anything could be accommodated for the phlogistic theory, Lavoisier's theory allows only one particular relation. The fraction that should be hydrogen (and released) versus the fraction that should be oxygen (and compounded with the metal) is known in advance from the synthesis of water from hydrogen and oxygen. Hence Lavoisier's enthusiasm in reporting that the inflammable air released would exactly combine with a weight of oxygen equal to the weight gained by the metal to re-form just the amount of water that had disappeared.

As with the puzzle of what happens during the combustion of a metal in pure oxygen (the "steel wool" experiment), this result can of course be quined. Taking the phlogistic view, we could say that the calx requires the same phlogiston content as the metal, so of course the amount of water absorbed must be in accord with that. To a sufficiently committed phlogistonist, this perhaps would seem another success for his theory. It has helped him discover yet another interesting fact, that the amount of phlogiston combined (as a component of water) in the calx is exactly the same as the amount in the metal. But to someone not committed to phlogiston, it is another example of the defenders' chasing after what Lavoisier's argument could foresee. It invites the reply that indeed the metal contains just the same amount of phlogiston before and after calcination, namely, zero.

Kirwan supported the view that inflammable air (hydrogen) is pure phlogiston. Cavendish, Scheele, and others provided variant views that gave an even larger role to hidden water. Cavendish, for example, argued that the *weight* of hydrogen and also the weight of oxygen, and apparently of gases in general, is due to water. The logic of this claim seems to require that not only all calxes but many other solids must contain substantial amounts of water (to provide weight for the gases released in various reactions).

I can't go very far into the highly complicated story of the varied accounts offered by defenders of phlogiston after 1783. The most detailed account I have found is written mainly to defend Cavendish's priority in discovering the composition of water (Wilson [1852] 1975). But phlogiston appears to have returned for Cavendish to its imponderable state, while something even more extreme held for oxygen. The weight of both gases seems to have become for Cavendish, as for Scheele, water in an aerial form. Oxygen (the element, not the gas) became even more ghostly than phlogiston.

Cavendish argued that Lavoisier's hydrogen is water enriched with phlogiston; water with a normal amount of phlogiston is water; water with a deficiency of phlogiston is oxygen gas. So oxygen (again, the element, not the gas) is, so to speak, the ghost of departed phlogiston.

During calcination the superphlogisticated water in the metal combines with the underphlogisticated water extracted from the air (oxygen gas) to form plain water within the calx. Priestley was moved to comment on how remarkable it is that water is a constituent of inflammable air and inflammable air is a constituent of water. But, on his view, that was the only possibility.[13]

Scheele suggested that elemental oxygen (that is, not the gas but the element that when combined with heat and water makes the gas) combines with elemental hydrogen to yield flame and heat, leaving water behind. Inflammable air, similarly, would be phlogiston plus heat (Lavoisier's caloric) plus water. So this was a pretty close variant of Cavendish's view, but independently arrived at.

On either of these views, where the weight of a gas is essentially the weight of the water it contains, one can explain why, when a calx is reduced with charcoal, the by-product is fixed air (carbon dioxide), while when inflammable air (hydrogen) is used, the by-product is water. On the phlogistic view, charcoal is made of phlogiston plus carbonic acid. The calx absorbs the phlogiston to revive the metal. The fixed air is released as a gas compounded from elemental fixed air plus the water from the calx, so that water provides the weight of (what we call) carbon dioxide, just as it provides the weight of inflammable air and of oxygen.

Kirwan, however, found Cavendish's view implausible. Who could believe, he asks, that hydrogen gas and oxygen gas are each already almost entirely water, even before they combine to form what we directly recognize as water? But his alternative defense of phlogiston requires belief that hydrogen and oxygen can form either water or fixed air. This resolves the by-product puzzle of the previous paragraph, though not in a way that Cavendish or Scheele found credible. On any of these phlogistic views, it remains an unexplained coincidence that what metals absorb when they give up phlogiston contains phlogiston— in fact contains just the amount of phlogiston that is given up.

Because no two defenders wholly agreed, it is difficult to summarize the response to Lavoisier's post-1783 argument. But as I have been indicating, in all versions there is now unobservable water in various places; in most, there is the oddity that inflammable air is a component of water and water is a component of inflammable air, which struck Priestley as strange even as he endorsed it; it is still a puzzle (in all the variants) that we never see a calx without its hidden water, and that we can never overtly extract the water except by adding light inflammable air (heating the calx with charcoal instead yields fixed air).

In Kirwan's version (as also in Priestley's), phlogiston plus oxygen only sometimes yields water, since the same ingredients can also yield fixed air (carbon dioxide). Both Kirwan and Priestley claimed experimental evidence for that. But the demonstrations they reported always are either hard to replicate or involve an amount of acid so small compared to the total amount of reactants that they can hardly explain away the dominant results. Nature stubbornly resists supplying the sort of evidence that would be good news for phlogiston and bad news for Lavoisier. Lavoisier's theory faces anomalies, complicated by conflicting experimental results. But the results that were inconsistent with his theory were always either marginal to his essential claims, or in dispute (someone reported X, but others got another result), so that none had the property of being a clearly demonstrated and *necessary* feature of the phlogistic view.

It is logically conceivable that the world could be as required for those ideas, or for even much more complicated ideas marked by even stranger coincidences, to be right. But pragmatically, avoidably complicated beliefs ordinarily cannot be sustained (stressing again that what is at issue here is an empirical point about what commonly occurs, not necessarily or directly a normative point about what we ought to endorse). Consequently, as Lavoisier's and his rivals' arguments become more familiar, so that a person comes to have an all-at-once sense of the whole network of beliefs required to make the phlogistonist view coherent, the gap between the economy of Lavoisier's scheme and the complexity required to sustain belief in phlogiston becomes more and more apparent.[14] It is *only* (I want to claim) in the light of a powerfully entrenched habit of mind that organizes the sense of what is going on in combustion and related processes in terms of the physical sense of something escaping—so that nothing that fails to satisfy that sense can feel right—that we can understand the evolution we have now reviewed: the virtual unanimity favoring phlogiston until 1783 (though logically the evidence was sharply conflicting and should have stimulated a crisis), and the continued resistance of many chemists for at least a few years even after the discovery of the composition of water gave Lavoisier a case that looks overwhelming.

5.5

We can now review the barrier claim in the light of this case. If a person had a powerful intuition that something is escaping during combustion (that is, if his network of beliefs were deeply entangled with the commonplace habit of mind with which we began this chapter), then the extent to which he might "quine" or "duhemize" experimental results could be very remarkable. To a person whose intuition is solidly

governed by that habit of mind, an account that gives no role to phlogiston can only be misleadingly simpler, since it is intuitively *wrong.* It
leaves a gross puzzle (where is the phlogiston?), parallel to the difficulty
that the first generation of astronomers confronted with Copernicus's
theory (where are the epicycles?).[15] Over time such commitments can
erode, though usually only when the individual is repeatedly confronted
with embarrassing counterarguments. So it easily happens that a person
no longer active in debating new research results never changes her
mind, even a person still active but so senior that it has become impolite
or impolitic or impractical to bluntly confront her.

What I will try to show in some detail as we proceed is this:
though it is always logically possible to quine any evidence, once we
allow for the existence of endemic cognitive propensities—the most
fundamental reflecting the shared Darwinian heritage of the species—
the possibilities for *effectively* quining in ways that can remain plausible
to third parties are severely restricted. It will be very hard for you or me
to give up a theory that seems a major accomplishment of our lives. But
it may become very hard for us to persuade anyone else that newer
arguments and evidence leave that theory still plausible.

Notice that, as suggested by the concluding discussion of chapter 1, peculiarly powerful habits of mind are marked by two characteristics. First, they are more than mere static definitions. We expect them
to be constructive, in the sense that we expect them to be tied to some
unfolding pattern of action or thought, often (as here) rooted in everyday experience. In the case at hand, the habit is tied to the physical
sense of what a person sees as a fire burns, the flames emerge and dissolve into the air, the fuel gradually disappears. But as a science develops, the physical experience out of which a deeply entrenched habit of
mind develops may move far from ordinary experience, and indeed may
come to be something that only specialists in the field experience in a
physical way—by manipulating apparatus, carrying out calculations,
drawing diagrams, making arguments that invoke physical analogies.

Second, the critical habit of mind will be *entangled* with multiple features of a person's experience: here by way of Stahl's synthesis
of combustion and calcination and with later broadening to include
links to respiration and fermentation. Note well that the entanglement is
not merely some verbal trick. Because phlogiston is effectively negative
oxygen, phlogistic arguments often led to striking results and worked to
unify a wide range of phenomena. But only a working chemist would
have been closely familiar with all that, and in particular only a chemist
would have *used* such thinking in routine practice and teaching. Hence
it makes sense that, as Lavoisier's writing makes very clear, nonchemists

in the Academy were far easier targets for his argument than were the chemists, and similarly that Priestley's nonchemist fellows in the Lunar Society were ready to abandon phlogiston on the basis of Lavoisier's 1777 argument, though no chemist of consequence was. In *Patterns* I give parallel illustrations from both the Copernican and Darwinian episodes.

Finally, it is pretty obvious in this case that the barrier indeed was *unique* (there is no competitor in sight that begins to account for the behavior we observe) and that it was mostly *tacit* (that is, having powerful effects on intuitions, shaping seeing-that even when it was playing no explicit role in the reasoning-why).

This case was selected in advance because it looked like a case that should yield interesting results from a barrier analysis.[16] And indeed, we get a story that diverges a good deal from the usual story, but in ways solidly based on the record, and that very neatly fits the barrier hypothesis. Detailed accounts of the episode naturally notice that things took a new turn after the discovery of the composition of water. No one who has looked into the primary record can miss that. But the water discovery has rarely been seen as crucial for Lavoisier's success. Conant even treats it as the basis of a *revival* of phlogistonist claims, allowing them finally to propose an answer [5.4] to Lavoisier's 1777 paper. Summary accounts—even those by writers of the best detailed accounts—commonly let this aspect drop from sight. But here water, not oxygen, is the key to the story, in the straightforward sense that, of the three discoverers of oxygen, two saw no reason to abandon phlogiston, and even Lavoisier, who saw that possibility, did not aggressively pursue it until after the discovery of the composition of water.

This much-studied case gives us (I would argue) a very clear example of how a precise sense of what separates individuals committed to rival paradigms can be given in terms of a particular, unique habit of mind, just as argued in the theoretical discussion [3.5]. Further, without an unreasonable stretch of usage, it seems to me that a person (like Cavendish or Priestley) on one side of the barrier that has been the focus of this account indeed lives in a different world than someone on the other side. What looked plausible to the defenders came to look ridiculous to Lavoisier. What looked obviously right to Lavoisier looked definitely wrong to the defenders even though they could not deny that it looked plausible. There is something going on that requires a special term—and Kuhn's "incommensurabilty" is the best that has come along—for characterizing this incompatibility, which cannot be directly breached by a logical argument, though ultimately arguments play an essential role in the social and psychological process that eventually moves a person from one side of the barrier to the other.

In this case, was the shift important? In contrast to the Copernican and Darwinian cases, the actual change in views looks logically like an episode in the development of pneumatic chemistry (with oxygen interpreted as negative phlogiston), not like a fundamental change of itself [4.5]. For such a case, is the shift, even if observably wrenching for the participants, usefully thought of as revolutionary?

In a logical sense it is reasonable to want to say no. Nevertheless, chemistry was handcuffed until the barrier was broken. The convoluted theories that Cavendish and other defenders were led to adopt after 1783 [5.4] led nowhere in terms of guiding research and thinking in fruitful ways. For those who crossed the barrier, striking new developments proceeded so rapidly (most conspicuously in the form of the radical reformation of the language of chemistry and of how to talk and think about chemical processes, with the model provided by Lavoisier's textbook) that it has often been claimed that Lavoisier succeeded *because* he trapped people with the new nomenclature and the new textbook. On the view here, such claims reverse cause and effect.

Six

The Emergence of Probability

Habits of mind, to repeat a fundamental point, are (like physical habits) by a wide margin functionally advantageous. Expert performance in every domain is contingent on well-entrenched habits, which allow most of what needs to be done to proceed fluently without conscious attention. By and large, habits function as *facilitators*.

Consequently, the barrier argument requires two variants: the barrier story as I have been dealing with it so far, and a converse story that I have only briefly mentioned, where an important new idea is cognitively difficult not because some barrier *blocks* our reaching it but because some odd move is required to reach it, and there is no habit in the repertoire that makes that available. Phlogiston provides an example of barrier effects. The long-delayed emergence-of-probability, which has remained puzzling despite the attention it has received in recent decades, gives us a case where the timing of an important and (logically) easily available discovery seems to turn on a missing habit. Until the habit appeared, a calculus of probability remained out of sight, though logically for upward of 2,000 years the idea lay waiting to be picked up by any clever individual with some mathematical competence.

The delay before an important idea was finally seen is even more striking for probability than for the Copernican case. No impediment here is comparable to the Copernican difficulty of seeing the Earth as a planet flying through space.[1] Only a handful of scholars alive at any one time were sufficiently expert in the details of Ptolemaic astronomy to have had a serious chance to make the Copernican discovery. But for probability, any mathematically inclined person familiar with symmetric dice knew everything a person logically needed to know. Symmetric dice date back at least to the second millennium B.C., and the mathematics required dates back at least to Euclid.

When a calculus of probability finally appeared, it did not come from some reasonably competent person who happened to be at the right place when a suitable concrete problem arose. Rather, the (to us) easy problem that ultimately proved particularly fruitful had been discussed for more than a century before Pascal, Fermat, and then Huygens were finally able to deal with it. Today even a reader with no training in probability theory should have little trouble seeing how the solution works. But in the seventeenth century it was extremely challenging. Fermat comments on the difficulty he has had explaining the argument to Roberval, the professor of mathematics at the University of Paris.[2] Those who could solve the problem in the mid-1650s seem to have been limited to the three most impressive figures then active in science and mathematics (Pascal, Fermat, and Huygens).

Now given that we reached the middle of the seventeenth century with the key idea still undiscovered, it is easy enough to notice what was going on, and judge that the emergence could no longer be very far off. During the several preceding generations, we see the long-delayed adoption of Hindu notation and the beginnings of algebra (both coming by way of Arab work). We then begin to see compact symbolism and (a distinct additional step, and more profound) the use of letters for indefinite numbers. We then get the development of analytic geometry and other steps that in a few more decades would lead to the infinitesimal calculus. Over the same period we have many new applications of mathematics to empirical problems, and especially the great impetus to mathematical analysis of nature provided by Kepler, Descartes, and Galileo. In the face of all that, it is not hard to give some account for the emergence of probability in the 1650s. What is a great puzzle is how such a simple, interesting, and useful idea (at a minimum, useful to gamblers, of whom there are always many) could have been missed for so many prior centuries.

6.2

On the barrier argument, we should be able to point to some plausible candidate either for a stubborn and strategic habit of mind that could block this discovery, or for some critical facilitating habit of mind that was missing. But the possibility of a barrier here is remote. If we think about endemic habits of mind that might be relevant to the emergence of probability (habits of mind entrenched by everyday experience in any culture), the puzzle of the long delay is not resolved but deepened.

No day goes by that a person does not have to make decisions that turn on which choice is more likely to have a good effect. Often the favorable outcome is fixed, so that the choice depends *only* on which move is more likely to get the favorable result. Further, many such choices involve more than one person, so that a common topic of discussion is what we should do, when the only issue is which choice is more likely to give us what we want. Hence an *explicit* comparative sense of probability as something with a greater or lesser character must have existed for thousands of years: for as long as our species was capable of discussion of cooperative choices in a risky world.[3]

Even primitive (and of course nonquantitative) intuitions of what we would call the weak law of large numbers seem pervasive. A medieval practice established the length used in measuring a piece of property by lining up foot-to-foot the first sixteen men to come out of church. Long before there was any formal notion of sampling theory, we see this crude means of limiting error by averaging over multiple observations. Loaded dice were uncovered in the ruins of Pompeii. And experience teaches everyone that if (say) there are many fish in pond X and only a few in pond Y, we will more often catch a fish if we try the pond with more fish—and the converse: if it is easier to catch fish in X than in Y, we readily believe there are more fish in X. Presumably, no one thinks such intuitions became common only after the 1650s.

Large-numbers intuitions show up most explicitly in the occasional enumerations of possible tosses with dice. In a paper from about 1620 (translated in David 1962), Galileo uses this device to respond to his patron's inquiry about why (with three dice) ten turns up slightly more often than nine.[4] Galileo responds with a listing of the 216 possible tosses, showing that 27 tosses yield ten, but only 25 yield nine. The key to his analysis is to allow for all permutations (not just 1,2,3, but also 1,3,2; 3,2,1; 3,1,2; 2,1,3; and 2,3,1), for in terms of combinations, ten and nine are equally available.

A point to note here concerns the way that Galileo explains why some numbers are thrown more frequently than others in dicing. The explanation is "very obvious . . . [it] depends on their being able to be made up with more variety of numbers" (David 1962, 192). Galileo is saying that, if a point can be tossed in more ways than another point, then in fact it will come up more often. Galileo does not present the argument as one that he needs to defend, or as one for which he claims any originality. Yet to formally deduce the "then" clause from the "if" clause amounts to a proof of the weak law of large numbers, which can only be done using the probability calculus. But intuitively the inference seems to be obvious to everyone, not only today when quantitative probability is familiar in ordinary language but long before that. The intuition is so readily prompted that it takes some work for a mathematically inexperienced person to see why there is anything to be proven.

So we would not expect that Galileo was the first to use this sort of analysis to account for what gamblers had observed about the behavior of dice. In fact, listings of the 216 possible outcomes to tosses of three dice can be found centuries earlier. But seeing permutations of the same combination (for two dice, 2,1 and 1,2) as different outcomes is itself counterintuitive, since in usual experience the two possibilities are indistinguishable. In elementary probability courses today, it takes some effort to make clear to students why permutations, not just combinations, must be counted. The most plausible explanation for why we find listings of the 216 permutations, not just the 56 combinations, is that Galileo's intuition—and that of earlier writers—was that it was obvious to them that the possible outcomes should be in proportion to the observed actual outcomes. Experience with dice taught that counting combinations gives wrong answers, so that (like Galileo) earlier writers were stimulated to think harder.

The several tabulations of permutations that survive have done so by accidental linkage with things preserved for other reasons. We have other such fortuitous clues to the presence of probability intuitions, such as a remark that comes down to us because it happened to appear in a commentary on Dante, or the unpublished memo Galileo prepared for his patron, which almost certainly would have been lost except for the great interest in anything Galileo happened to write (David 1962, 61–69). The same fortuitous character holds for a few pages on calculating chances with dice included in what is otherwise a guide for novices provided by Cardano. Cardano never published this material and mentions it only in passing in his autobiography, which certainly is not otherwise shy about claiming originality. Such material by less famous persons or not tied to material preserved for other reasons, would be lost. So it is a reasonable presumption that the surviving examples of early probability analysis are only a small sample of what a complete record would show.

A striking point about what has reached us is that we never see the insights defended in detail, as we would expect for a claim the writer expects to be surprising to his readers; nor are the ideas presented with any suggestion that the author would like to be credited as the first to point to the result. Rather, elementary probability intuitions are treated as what everyone knows or takes as obvious. Galileo does not explicitly claim originality for showing the 216 permutations. Although he is clearly pleased with his paper, as he had good reason to be, there is no claim that he is the first to work out this analysis, a claim that he is certainly not shy about making on many other matters. Galileo makes no claim of novelty in using the number of ways a number can be tossed as a way to measure "very accurately all the advantages, however small" in dicing. From our point of view it appears that probability is not only close at hand, but actually being used. But there is no indication that those involved see what they are doing as a fundamental innovation, or indeed as doing anything that goes particularly beyond common sense. With the very doubtful exception of Cardano, no general methods of probability analysis emerge, only enumerations of some special case.[5]

6.3

What crucially distinguishes all this precursor material from what emerged around 1655 is that there is no indication that anyone saw the results of such enumerations as quantities, which could then be used as inputs to further calculations to assess the probability of sequences or combinations of these cases. Put another way, no one saw the beginnings of a calculus of probability. Rather, the particular enumerations were seen as a way of explaining comparative probability insights or observations already in hand. But they implicitly reveal the presence of comparative probability intuitions much like ones that are common today. Examples of what we today call Pascal's triangle, which would most plausibly arise out of consideration of permutations and combinations, go back much further than the dicing enumerations. But no clear case survives of successful multistep combinations, meaning those that involve sequences or combinations of one-step combinations.

If the argument here is right, no culture should be found that entirely lacked some functional analogue of basic probability notions, though the manner of framing probability intuitions (for example, in terms of what the gods will do) may look strange to us. But also no case would be found before about 1650 that goes clearly beyond these endemic intuitions. I comment in note 5 on why Cardano's guide—dated about a century earlier—is not a convincing counterexample.

6.4

If in fact comparative probability intuitions are endemic, then a *barrier* to the emergence of probability is unlikely. But a candidate for a missing facilitator readily presents itself. The very slow emergence in general of novel extensions of the notion of number—a very marked feature of the history of mathematics, with the teasingly slow acceptance of zero as a number providing an exemplar—suggests that the tardiness of the emergence of probability was itself an instance of the broader problem.

Once looked at in this broader framework, a salient suggestion is that the natural notion about quantity—the only notion that extends far enough back so that it has no history, since it seems to have been always around—is of "number" as tied to something you can count *to* using the intrinsic lumpiness of the world, or at least lumps familiar for so long that no one has a sense that they could be lacking. Apples, days, and so on are examples of the first. Bushels and dollars provide examples of the latter.

Although counting *to* a number is endemic, there would be no endemic habit of mind that prompts us to the inverse process of seeing a number as something a person could construct, given a quality that invites "greater or lesser" comparisons. In that inverse process, we are not seeing lumps and counting them. We are seeing something that looks quantitative and figuring out or defining lumps to count it with. That inverse process—what I will call inverse counting—could only become common experience, hence the basis for a habit of mind, once examples of numbers seen as constructions were already often encountered. So there is a chicken-and-egg quality to this puzzle. Inverse counting becomes intuitive only after it becomes familiar enough to begin to establish a habit of mind that prompts a person to look for some way to quantify qualities that do not naturally come in already familiar, directly countable physical lumps.

But by the middle of the seventeenth century, among people fascinated by the new scientific ideas, such numbers begin to appear, driven by work on concrete problems that (notably) Harvey, Galileo, and Torricelli had attacked with striking effect.

Pascal was both a cofounder of quantitative probability and the man who carried on (with memorable results) Torricelli's discovery that we live at the bottom of an "ocean of air." Showing how the weight of the atmosphere could be quantified was one of the first really striking examples of inverse counting. The length of a column of fluid supported by the atmosphere came to be interpreted as a measure of the weight of an imagined column of air of the same area extending from the surface of the Earth to the top of the atmosphere.

Early in his *Two New Sciences* (1638), Galileo speaks of speed as a matter of degree, and elsewhere he speaks of the thermometer he devised as conveniently marked off in degrees. But (most of the time, at least) these degrees have only ordinal significance. A speed of 8 degrees was faster than a speed of 4 degrees, but not necessarily just twice as fast.[6] Galileo only gradually starts to talk of speed as if a single number could express this combination of distance and time, and he never explicitly uses such constructed numbers.[7]

On the argument I want to make, seeing instantaneous velocity as something that could be captured with numbers—even if at first only by way of a proportion—is critical. The further difficulty of coming to feel comfortable with numbers of mixed dimensions (such as miles per hour) is not, I think, so central an issue for the emergence of ideas that use such numbers. So I will not argue that the emergence of probability turns on whether the chance of getting two heads in a row is thought of as a proportion 1 : 4 or as a single number 1/4. On the argument here it is a further step, not cognitively easy, but soon forced by the clumsiness of doing calculations without it, to go from the proportion to the single number. The cognitively harder though logically easier step is seeing that some hitherto only comparative notion can be made quantitative—even if at first only as a proportion between directly countable numbers.

Today inverse counting is wholly routine. For almost anything that leads to comparative greater or lesser judgments, we are tempted to look for a way to attach a number. We do that for subjective probabilities, for the performance level of chess or tennis players, for temperature, for temperature adjusted to allow for humidity, and so on. We are more likely to blunder by taking seriously meaningless numbers than by failing to realize that a number might usefully be attached to some hitherto only comparative variable. We find it almost as familiar to construct numbers as to count to them, though of course the former is far less common and something that only gradually becomes familiar, not something learned on "Sesame Street." A child learns to count and then (usually with some difficulty) learns to accept various sorts of numbers you can't just count to: zero, negative numbers, the abstract numbers of algebra, imaginary numbers.[8] Gradually, attaching a number to anything that invites or requires greater or lesser comparisons becomes wholly familiar: it has become what we *habitually* expect.

6.5

The line between direct and inverse counting is context-dependent because with taming (this is a [benign] example of taming),[9] once you have become familiar with how to count to a number, what might on first sight seem an unreasonable and artificial device comes to be seen as an ingenious way of seeing how a number can be pinned down. We first see puzzlement about how that would work, or in a more self-confident novice incredulity that there is any well-defined number. With familiarity in working with the number (this requires that in fact the number works well enough to do things with it), a person loses any doubt that the number is there, but for a while retains a sense of cleverness in seeing how to reach it. With still more familiarity, that wears off as well: both the quantitative character of what had been only a comparative value *and* the sense of ingenuity in seeing how to count up that number fade. No one today sees anything clever in using a simple number to give us miles per hour, batting averages, temperature, and so on.

Extensions of direct counting date back a long time, for example (for astronomers), angular separations as measured by equally spaced markings on an observing instrument, and (for everyone) the use of artificial measures of exchange value, growing out of what was originally unmediated direct counting of some commodity of compact value like beads. Such extensions of direct counting, repeating that important point, are so familiar and of such long standing that they occasion no sense of attaching a number to something that anyone would imagine was anything but intrinsically countable in terms of some perfectly familiar discrete units. We grow up familiar with money and prices. A person is no more taught about that than she is taught to speak her native language. It is just there all the time. The same is true of many other examples of inverse counting, including quantitative probability itself in modern societies, where such notions are part of ordinary language.

As the number of cases of constructed numbers increases—in particular cases of constructed numbers that a person has to be taught to comprehend (in contrast to, say, money prices), then some intuitive sense of inverse counting as itself a familiar process must start to emerge. As the number of such examples increases, a person inevitably encounters some such numbers under conditions where that initial sense of doubt when the notion was brand new would be prompted for that person, and would have to be eased away by instruction and experience. The teacher, perhaps more than the student here, would be prompted to awareness that a process had to be invented to allow counting to this number, but with experience would start to find that procedure natural and intuitive, the way a child after a while finds that riding a bike is no longer challenging but effortless.

In the decades preceding the emergence of probability, a series of striking examples of inverse counting came too close together to be wholly tamed, as essential features of a wider class of examples of new applications of mathematical reasoning to empirical problems and within mathematics itself.

The first example that seems to conspicuously involve inverse counting seems to be Galileo's early work on what we recognize as instantaneous velocity. Harvey's calculations of the rate of flow of blood came a little later, though they were published a little earlier, and provide a more ambiguous example. Soon after Galileo's *Two New Sciences* (1638), we have the further very influential example of inverse counting in Torricelli's proposal for measuring air pressure (1643) by the length of a column of mercury it will support, then Pascal's brilliant extension of this line of inquiry. And of course as background we have the commitment of the dominant scientific figures of the preceding decades (Descartes, Galileo, and Kepler) to mathematical analysis of nature, which in turn was powerfully shaped by the astonishing result produced by Copernicus from an unabashedly mathematical viewpoint.[10]

So by the middle of the century following Copernicus, it is not surprising that new habits of mind would be developing that made inverse counting a pragmatic possibility, and that indeed had started to make that tendency habitual among those most intensely involved in the emerging scientific enterprise.

6.6

Once things were ripe for the emergence of probability, it appeared quickly. But how it emerged may look backward to a modern reader, for whom odds and expected value derive from probabilities. But those notions start as special sorts of prices, and probability emerges from them—odds far back in prehistory, and what we would call expected value in discussion of risky contracts in the commercial world of the Renaissance.

Since gambling is pervasive across human cultures, occasions must arise in which I would like to make a bet but you are reluctant to take it. What does it take for me to tempt you to bet? The obvious move is for the person who wants to bet to offer to put more money in the pot, as the obvious move for a person who wants to tempt me into trading some beans I have for some corn he has is to offer to throw more corn into the deal. Betting at odds emerges far back in prehistory, and although that must be tied to endemic probability intuitions [6.2], it no more requires quantitative probability than prices require quantitative utility values.[11] It is one of the many things we can do without having a theory of how we do it, and even without having a glimmer of the possibility that there might be such a theory. When theory first emerges, among its first results will be an explanation of what might lie behind what people have been doing for a long time without the help of any theory.

Somewhere along with the notion of price comes the notion of fair price.[12] This too seems to have a long prehistory. Extended interest in the notion of fair price appeared with the development of commercial activity in the Renaissance, leading to a good deal of discussion of how contractual arrangements should be adjusted for risk of default (so rationalizing departures from the church's prohibition of usury), or piracy, and so on. These commercial questions seem to have slipped over into questions about gambling, where analogous "fair price" questions could arise, but in a context where much more exact experience about risks was available.

We can see at work the primitive probability intuitions discussed earlier. Though without the explicit laying out of assumptions we expect in a modern discussion, it is pretty clear that people think the fair price in a risky context is the price that over the long run would give no special advantage to either side of the bargain: what we would call today the expected value. Here the striking point is that, for a long time (on the order of two centuries), simple gambling analogues of this problem were discussed, but no one could solve them. We find it hard to avoid seeing odds and probability as pretty much the same thing, but our predecessors took a long time to get from the familiar notion of odds, or from a notion of fair price in the context of risk, to the unprecedentedly novel notion of probability as a number between zero and one. What we need to look for to pin down the emergence of probability is the passage beyond the primordial sense of odds as a kind of price to a sense that odds could be calculated, from which the sense of probability as a number between zero and one could develop. Since the record is incomplete (we pick up the Fermat-Pascal story when both have already gone a considerable way), we can conjecture several stories broadly consistent with the inverse counting argument, of which I will develop the version that seems to me the most plausible.[13]

6.7

Gambling analogues of the fair-price problem are convenient, since here possible outcomes are cleanly defined, and (as the enumeration of possibilities for throwing dice illustrates [6.2]) the results of counting possibilities prompt ready intuitions about how to judge relative advantage. A particularly suitable puzzle to prompt seeing fair prices for gambles as something that might be calculated would obviously be where a quantitative result is required (not just a comparative one). Less obviously, the most fruitful sort of problem would be where taking what appears at first sight to be the balance of favorable to unfavorable cases gives a noticeably *wrong* answer. For it is in just that sort of case where more explicit work on how to count cases in a way that works would be stimulated, and the result might be recognized as a novel and generalizable extension of quantitative thinking, not just as a special case that yields a solution of no wider interest. What seems to be needed, then, is a context in which a person has to *contrive* a way to count cases, so that the intuition could be prompted that in other cases as well it might be possible to contrive ways to count to probability numbers in ways that are not immediately apparent.

We might have supposed that analyzing dicing possibilities in terms of possible permutations, rather than combinations, would have been exactly what was needed [6.2]. But on the record it was not sufficient. We know that was done repeatedly, but no step beyond it followed. A possible reason has already been considered. Inveterate dicing was a vastly more widespread recreation in earlier centuries than it is now. From such experience a person could know that just counting combinations gives the wrong answer—since that sort of count sometimes implies equal probability for pairs of possibilities that any experienced dicer knows are not equal possibilities. Extensive experience with dicing also implies occasional experience with tossing unmatched dice. But if the dice are unmatched (say red and white, or smaller and larger), then players see permutations directly. Indeed, if in addition to dicing in general being more common, the use of dice that were not of a matched set was less unusual in earlier centuries than it is now, it could be that the distinction between permutations and combinations, which a student today needs some work to catch, might then have been a ready intuition from common experience.[14]

On that conjecture—and on the record, so perhaps that conjecture is right—seeing that one must count permutations did not provide the stimulus needed to yield a generalizable experience of calculating odds. Instead, the problem that turned out to play this role was what was called "the problem of points."

6.8

What dominates accounts of the emergence of probability is a particular form of the problem of points that had been posed to Pascal by the chevalier de Mere. But Pascal mentions that the problems come in two forms (FP 230). Pascal and Fermat wrote mainly about one, since de Mere himself could solve at least simple examples of the other version and naturally Fermat and Pascal mainly discuss what de Mere could not solve. This is worth attention, since logically there is actually no difference between what I will call "points with players" and "points with dice." But a cognitive difference exists between the two that fits the story I am sketching, since as things went, points with players turned out to require a more artificial method of counting to reach the right answer than did points with dice.

The problem of points with dice is to calculate where the advantage lies in bets where a player has a certain number of chances to make a given throw. Apparently de Mere (and also Roberval) worked out the case where four throws are allowed to roll a given point with a single die.[15] We know about this because de Mere used this result to argue to Pascal that mathematics could not be trusted. He reasoned that since four throws were sufficient to make it a favorable bet to throw one of the six points with one die, then since $4:6 = 24:36$, 24 ought to be sufficient for making a particular one of the 36 possible outcomes with two dice. Apparently working out an exact answer for the two-dice case was too much for de Mere. But from his experience with dice de Mere assured Pascal that it was best to bet against making a twelve (that is, 6,6) in 24 throws, although his mathematical extrapolation said the contrary. So mathematics could not be trusted.

Pascal recounts de Mere's argument with the comment that de Mere is a clever fellow but no mathematician. Pascal takes it for granted that Fermat sees where de Mere made his mistake (he gives no explanation), so perhaps this problem of points with dice was discussed before the first letter that survives, which is clearly not the first in the series. But we can see that de Mere apparently knew enough—and therefore we can surmise quite a few people by now knew enough—to work out that $5 \times 5 \times 5 \times 5 / 6 \times 6 \times 6 \times 6$ was the chance of losing with four chances to win, which is less than half.[16] But working $35^{24} / 36^{24}$ was more than de Mere cared to try, so he took his reasonable-sounding but incorrect shortcut. Fermat and Pascal, however, were no doubt familiar with logarithms (invented much earlier in the century), so that the second problem would have been simple as well for them, and it yields a result consistent with gamblers' experience.[17]

At this point the incompleteness of the record—in particular our ignorance of how much discussion between Pascal and Fermat preceded the first letter that survives—leaves two salient possibilities. One is that dealing with the problem of points with dice in fact is the breakthrough to a calculus of probability, which had happened before the first letter we have. Indeed, since Pascal mentions that (at least) de Mere and

Roberval had already solved the computationally simpler form of the points with dice problem, then perhaps the apparently critical role of Pascal and Fermat dealing with the problem of points with players is only an artifact of the incomplete record. But the very fact that de Mere (and more significantly Roberval, who was a leading mathematician) could not deal with the problem of points with players persuasively suggests the contrary.

Even if the problem of points with players played a special role, quantitative analysis was by then so readily being seen as relevant to questions about gambles that the emergence of probability could not be far off. The points with players problem, discussed for many decades, finally became one that could be solved, although it took a Fermat or Pascal or Huygens to do it. The key both to the difficulty and (on the argument here) also to the fruitfulness of this problem is that it is hard to see as one that can be solved by just counting cases. Further, the answer being sought is not just a comparative answer (which side of the bet is advantageous). A quantitative answer is needed, since the points with players problem turns on how much a position is worth when a match is not yet completed. The answer must be something more than zero (since there is a chance of winning) and less than the full stake (since there is a chance of losing).

6.9

Here is the problem. Suppose a prize will go to the first player to win a certain number of points in some fair game. And suppose the game is interrupted before the required number is reached. How should the prize be divided?[18] Sometimes the problem was posed in terms of a fair price for a third party who wants to take over a player's position. Although numerous efforts were made over the century and more preceding the 1650s, no one seems to have been able to solve what to a modern reader is a trivial problem: which in fact is not logically a different problem from the points with dice problem that de Mere and Roberval were able to solve.

Consider a particular case where we are tossing coins, with heads a point for me, tails a point for you, and the winner to be the first to reach five points. How large a share of the prize should go to me if the game is interrupted when I have four points and you have two? Or, to strip the matter to its essentials, what is the fair division if you need three points before I get one?

The simplest count (with "H" a point for me, "T" a point for you) would say I win for the cases H; T,H; T,T,H. You win for the case T,T,T. On this count my ratio of favorable to unfavorable cases is 3 : 1.[19] Alternatively, I could count as if we were going to play as many points as would be required for the trailing player to win, ignoring the fact that in an actual situation play would stop as soon as the issue had been decided (as the World Series stops as soon as one team has won four games). Counting in this way the possibilities are H,T,T; H,H,T; H,H,H; H,T,H; T,H,T; T,H,H; T,T,H; T,T,T.

Of these eight possibilities, the trailing player, as before, wins only one, but now there are seven cases where he loses instead of three. So there is a straightforward way to count that indicates a fair division of 3 : 1 and an artifactual way to count that gives the fair division as 7 : 1. The critical step in the emergence of probability (as revealed by the Fermat-Pascal correspondence) was seeing the artifactual possibility, and realizing that it alone gave the right answer. Fermat and Pascal had trouble explaining the reasoning to what we must suppose was an exceptionally qualified person, the professor of mathematics at the University of Paris, Roberval.[20] Indeed, even for a sufficiently naive modern reader, it might seem that what Fermat and Pascal claimed was the right division to the problem as posed was changing the problem.

Yet the procedure that implies a 7 : 1 division gives no advantage to the leading player, who so plainly gains an advantage in the apparent result. The trailing player loses nothing whatever by agreeing to let play proceed even if he has already lost. He still wins in the one case where he would win anyway, and loses only in cases where he would lose anyway. So a fair division cannot actually be different for the normal case where play terminates when a decision is reached versus the artifactual case where play continues though it cannot affect the result. Nor, once attention is drawn to the point, can anyone doubt that it is easier to toss a single head than a string of three tails, so that just counting cases in the original way cannot give the right division.

Once the two possibilities (3 : 1 versus 7 : 1) are explicitly on the table, close scrutiny would be stimulated to clear up the conflict between the two. A good deal of the Fermat-Pascal correspondence consists of exchanges of proposals about various ways to analyze the issue, working this out in terms of concrete cases instantiating various versions of the problem. Fermat and Pascal satisfied themselves, and before long satisfied Roberval and others initially skeptical, that indeed the right way to count includes cases that would never actually occur, like H,T,T in the example used here.

In general, once you have seen the right answer is 7 : 1, because the leading player can expect to win seven times out of eight such encounters, then you have a powerful aid to seeing a far simpler way to get that answer, and one with very wide potential for application to other problems: sooner or later someone will see that the only way for the trailing player to win is for the coin to turn up tails three times in a row, and notice that $\frac{1}{2} \times \frac{1}{2} \times \frac{1}{2} = \frac{1}{8}$, which has a salient kinship to what you already know is the right answer. In particular, since it is natural to work out (as Pascal and Fermat did) the result for a range of cases (where the trailing player needs 2, 3, 4, . . . , and so on wins), the sequence of powers of two could hardly be missed.

How could it be that Roberval and de Mere understood how to deal with at least simple cases of points with dice, but had trouble following Pascal's reasoning about a points with players problem, which was simpler still?[21] The key clue here is that even Fermat and Pascal do not see that there is really only one problem of points, not two distinct (with players, with dice) problems. Apparently no one *before* the emergence of probability saw the players problem as just a trivially different variant of the dice problem.

The way I have treated the question here (as a problem in coin tossing), points with players is *transparently* just a simpler instance of points with dice, where the gamble is that a coin will be tossed x times without a tail turning up, just analogous to the problem of tossing a die x times without a six turning up.

The difference in the framing here from what Pascal and Fermat dealt with, hence what we can surmise holds the key to an explanation, is that in the seventeenth-century framing nothing is discussed but some generalized game to be won or lost. Implicitly, the situation is treated as a fair one in which (we would say) the probability of either player winning the next game is .5. So to us it is immediately intuitive that the game is like coin tossing. But on the record, until the emergence of probability—and of course we are here trying to understand what was going on just prior to that—it was *not* immediately intuitive to see chances as quantitative, and hence to see the chance of winning the next point in an unspecified contest (perhaps a game of ball or of chess) in a way that made it natural to compare chances across contexts where a comparative choice was not being faced, in the way that it would have been perfectly natural to compare prices of different items even though no exchange was at issue. So (on the record) it was not readily intuitive that the chance of winning a fair game is the same thing as the chance of winning a coin toss, or of rolling an even point with a single roll of a die. That is not at all a logical point. Logically, the coin toss is just an example of a fair game. However, not only Roberval and de Mere but also Pascal and Fermat (who managed to solve the problem nevertheless) did not see what is trivially obvious to later actors.

When de Mere correctly calculates the winning fraction in the dice problem as 671 : 625, he *is* counting cases that would never occur if we toss a die repeatedly to see if we get a six within four throws (FP 235). Yet he and Roberval do not see that the same needs to be done in the points with players context, and Roberval (at least, as we know from a comment by Fermat [FP 247]) needed convincing to believe it even after being told. It is hard to see why that could be, other than by noticing another habit-of-mind effect—distinctly secondary in the overall story, but providing a possible explanation of details of that story. Our experience in the players context is always that we stop when a winner is determined. Once one team has won four games of the World Series, the series is over; and the same for all analogous situations that we know and that seventeenth-century people knew. But for the dice case, the usual way to proceed would presumably have been to roll four dice at once and see if there is a six, not to roll one die up to four times.

Illustrating a fundamental point that has come up already, *logically* it is obvious that the chance of winning, hence a fair division of stakes or a correct assessment of where the advantage lay, cannot depend on whether we throw four dice at once, or one die four times, or one die up to four times but stop (since the gamble is decided) at less than four if a six has already turned up. So logically, it is a complete mystery why the dice and players problems were not seen as trivial variants. But entrenched experience in the world made seeing the dice problem as one in which it was natural to count all possible sets of four tosses (since ordinarily four dice would be tossed together), and not so the very same problem framed in terms of players (who would always be playing games sequentially, stopping once the outcome was certain).

On the other hand, that even de Mere and Roberval were able to solve the dice problem, as (arguably)[22] no one earlier had done, shows how essentially inevitable the imminent emergence of probability had become by the 1650s, and how logically small the further steps taken by Fermat and Pascal, then by Huygens, had become.

6.10

What the Fermat-Pascal correspondence reveals is the two most distinguished mathematicians of the age carefully working their way toward this (for us) quite trivial insight *and not quite getting there*. Pascal and Fermat never get beyond calculations of odds, though they are now dealing with cases sufficiently complex that the long-delayed explicit emergence of probability cannot be far off. They do get beyond the simplest case, where a single win by one side is needed against some multiple of wins by the other. But they get to the more complicated cases by a kind of induction that involves no further conceptual novelty.[23] Within two years, however, Huygens—who had heard of Fermat's and Pascal's interest in the problem, and (very helpful) of reports that they had solved it—reworked the problems and soon published (1657) what we can recognize as unmistakably a calculus of probability.

But even Huygens's little guide to the subject always speaks in terms of what we call, and Huygens called, *expectations.* He never explicitly goes beyond quantities that can be counted to in a straightforward way: the number of winning cases, the number of all possible cases, and the value of the stake. Results are always a directly countable amount of money, obtained by multiplying the total value at stake times the proportion of favorable cases. We recognize the latter as a probability. Apparently, Fermat, Pascal, and Huygens did not quite do that.

Even half a century later, the idea that probability can be a number is still counterintuitive. The introduction to an English edition of Huygens, written in 1715, tells its readers: "To reduce the inconstant and irregular Proceedings of blind *Fortune* to certain Rules and Limits, and to set a definite *Value* upon her capricious *Favors* and *Smiles* seem to be Undertakings of so chimerical a Nature, that there is no Body but must be delightfully surprised with that Art which discovers them both really possible" (W. Browne, in Huygens 1715 translation).

6.11

Summing up: Until it can occur to someone that you might figure out a way, and find it highly fruitful, to count even though there is nothing directly confronting you to count: that is, until what I have been calling inverse counting emerges from work on concrete problems (involving instantaneous velocity, circulation of the blood, air pressure, and some others), quantitative probability apparently cannot emerge— certainly does not emerge. Reaching a quantitative general notion of probability requires attaching a number to something that, unlike cows or dollars, you don't routinely encounter as something countable.[24] But everything logically needed had been available at least since Eudoxus's theory of proportions was developed 2,000 years earlier.

The advantage of starting from dice problems is akin to the advantage that the new science in general exploited in starting from astronomy. We get clean cases that facilitate the breakthrough analysis: for physics, uncomplicated by resistance of the medium for the motions of the planets; for probability, uncomplicated by anything to significantly challenge presumptions of equally easy outcomes for symmetric dice.

But once the habit of mind develops that prompts seeing the possibility of usefully attaching numbers to comparative values even when there is nothing immediately apparent to count (so that a person can now also think that if something looks comparative, then perhaps a good way might be *worked out* to count something that will measure it and make it quantitative), then quantitative probability was finally within reach.

6.12

A further puzzle concerns why, once probability emerged in connection with dicing problems (what Hacking calls aleatory probability), it was so quickly followed by generalization to cases in which there is no physical setup that gives equally likely cases, or where there is no long sequence of repeated cases, leaving us only with what Hacking calls epistemic probability. On the habits-of-mind argument, epistemic quickly follows aleatory (or stochastic or statistical) probability because, in experience, it is actually *prior,* with aleatory probability a special case of epistemic. That is what the discussion [6.2] of endemic probability intuitions is about, which is the point on which the account here diverges most conspicuously from Hacking's.

On the account here, once probability finally emerges from that background of endemic comparative probability intuitions, quick generalization comes very naturally. Habits of mind endemically entrenched by experience of life in a risky world would favor that. So in contrast to the more than 2,000-year wait for quantitative probability to appear, we here get a good case of a discovery (extension of aleatory probability to epistemic) that logically seems bold and even startling but that came very quickly, since already entrenched habits of mind facilitated it.

What I eventually suggest (in chapter 12) is that a parallel story suggests a general interpretation of the Scientific Revolution, where the inverse counting stressed in this chapter is itself put within reach by a kind of inverse habit of observation that emerges under the pressure of reconciling a Copernican world with commonsense perceptions of a stable Earth.

Seven

A Ptolemaic Tutorial

The next several chapters extend the Copernican analysis that provided the principal illustrations of the effects of habits of mind in *Patterns*. As background, I start with a general account of how the Ptolemaic system works. A reader who wants to acquire confidence in the technical issues should read with a pencil in hand, sketching the diagrams. But I have tried to make the subsequent chapters manageable for a reader who has only a general sense of how things work.

Within this chapter, section 7.2 lays out the main features of planetary movements. Section 7.3 sketches how the Ptolemaic models capture all these features in an elegant and (in its basic structure) remarkably simple scheme. Section 7.4 gives a general overview of the Ptolemaic cosmology that results. Section 7.5 then introduces three moves—each quite simple (certainly within the reach of a good student of eighth-grade geometry), any one of which allows a Ptolemaic model to be transformed into a new geometrical configuration that leaves all planetary observations exactly as before. Finally, section 7.6 shows how the moves can be combined to transform the Ptolemaic models into observationally identical (within the limits of naked-eye astronomy) Copernican and Tychonic models.

7.2

Someone who systematically watches the heavens will come to notice five wandering stars (Mercury, Venus, Mars, Jupiter, and Saturn) that move against the background of the fixed stars. As days pass, the Sun, the Moon, and these wandering stars (planets) move in a generally west-to-east direction against the background of fixed stars, and against the vastly more rapid east-to-west daily rotation of the heavens. The movement of each planet turns out to be tied to the movement of the Sun, in a double sense. Each planet moves always within a narrow belt (the zodiac) bordering the path of the Sun (the ecliptic). And the motion of each planet is tied in some way to the position of the Sun relative to the Earth.

For two planets (Venus and Mercury) the connection is particularly simple. They always stay close to the Sun. Since they accompany the Sun, their basic period (the mean time to return to a given position among the fixed stars) is necessarily just one year. The other three planets (Mars, Jupiter, and Saturn) move independently of the Sun, with periods of two, twelve, and thirty years. The location of the Sun gives no clue to where within the zodiac these planets will be found. But in another way the motions of Mars, Jupiter, and Saturn still turn out to be precisely coordinated with the movement of the Sun. Each of these planets exhibits a periodic reversal of the west-to-east motion, producing over a sequence of nights a backward looping against the background of fixed stars (figure 7.1). These *retrogressions* always and only occur during periods when the planet is high in the sky at midnight, or (roughly) when the planet is rising as the Sun is setting.

So retrogressions by Mars, Jupiter, or Saturn occur when the Sun is on the opposite side of the Earth from the planet, or, as astronomers say, retrogressions occur when a planet is in opposition to the Sun. Figure 7.1 shows a sequence of retrogressions, as they would be determined by an ancient observer who plotted the position of the planet against the background of fixed stars over the dates shown.

Venus and Mercury also exhibit retrogressions. But they are not easy to observe, since these planets are always close to the Sun. What is most noticeable for these planets is that they cycle in a slightly irregular way (comparable to the irregularities seen in the plots of retrogressions in figure 7.1) from a position ahead of the Sun to a position trailing the Sun.[1] After reaching some maximum position ahead of the Sun (28 degrees, at most, for Mercury; almost 48 degrees for Venus), the planet *slows,* is overtaken by the Sun, and returns to a trailing position.

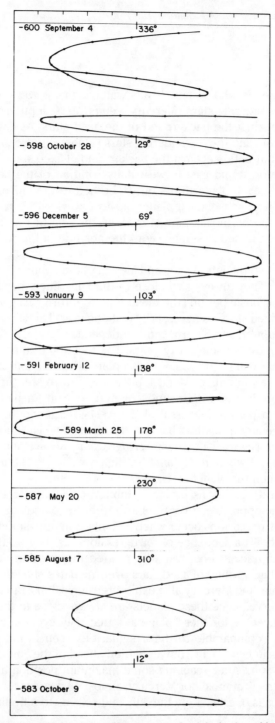

Fig. 7.1. Retrogressions of the planets. Here is a sample of the backward looping motions (retrogressions) traced by a planet (Mars, on the dates shown) as it moves through the background of fixed stars. The size and shape of the retrogressions vary, as do the intervals between retrogressions. The variations are around some mean (average) size and period characteristic of the planet. (Printed with permission from Owen Gingerich.)

The time for a total cycle (a fast phase followed by a slow phase, returning to the same position relative to the Sun) is constant for each planet: about 122 days for Mercury; about 584 days for Venus. But the phases are not equal: the fast phase takes longer than the slow. For Mercury the difference is modest. For Venus it is very striking: 440 days for the fast phase; only 144 days for the slow. Overall, there is a puzzle of why the planets divide into two sets; for Mars, Jupiter, and Saturn, a salient particular puzzle is accounting for why they perform their looping motions and for the variations of the loopings; and for Mercury and Venus a salient puzzle is accounting for their cycles, and in particular for why one half of their cycle takes longer than the other half.

A further puzzle is why retrogressions are much larger for Mars than for Jupiter, which in turn exhibits larger retrogressions than Saturn. Each also grows brighter as the planet reaches the midpoint of a retrogression, with the strongest effect for Mars and the least for Saturn. Mercury and Venus, in contrast, show no substantial variation in brightness during retrogression. This in fact gives an observational clue to the Copernican nature of the world, but was never so interpreted (even by Copernicans) until after the telescope could provide more direct evidence.[2]

As can be seen in figure 7.1, the retrogressions vary in a complicated way from one cycle to another in terms of their shape, their spacing, and their apparent size (all relative to the background of fixed stars). But sustained observation shows these are variations around a mean, not random. And the timing with respect to the relation between the position of the planet and the position of the Sun is absolutely reliable. The retrogressions for Mars, Jupiter, and Saturn, as already stressed, invariably occur and only occur when the planet is in opposition to the Sun. The timing of parallel effects for Mercury and Venus are also invariable and also occur when the planet is in line with the Earth and Sun, though these planets are never in opposition to the Sun.

For centuries astronomers pondered these matters, accumulating records of observations. The solution to the puzzles that Ptolemy finally produced, drawing on earlier ideas but adding crucial points of his own, turned out to be decisive. As I will try to bring out, the Copernican system was discovered not by abandoning Ptolemy but by building on what he did.

7.3

Ptolemaic astronomy is built around the use of epicycles: planets do not directly move on an orbit but are carried around on a sec-

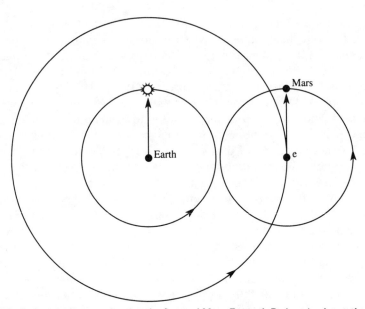

Fig. 7.2. Ptolemaic diagram showing the Sun and Mars. For each Ptolemaic planet, there are two orbits (the s-orbit and p-orbit defined in the text), one of which rides on the other: for Mars, the s-orbit rides on the p-orbit. The arrows show the common direction of rotation for the epicycle (literally, the circle removed from the center), the deferent (which shows the motion of the center of the epicycle, e), and the Sun (which in a Ptolemaic diagram of course orbits the Earth). At all times, and for every planet, the radius of one of the orbits (for Mars, Jupiter, and Saturn, the epicycle) lies parallel to a radius pointing from the Earth to the Sun. In the text, that is called the s-orbit (orbit linked to the Sun). The p-orbit (orbit carrying the intrinsic period of the planet) turns at a rate peculiar to the particular planet. Figure 7.5 and its accompanying discussion introduce the Ptolemaic cosmology behind this diagram. References in the text to compass directions in these figures are conventional, not literal. Since the Sun is shown due "north" (that is, directly up as the figure is drawn) from the Earth, Mars must be due north of the center of its epicycle, and the s-orbit of every other planet must point the same way.

ondary orbit, whose center in turn is carried around the primary orbit (figure 7.2). This idea turns out to work so well that Ptolemy and later astronomers (including Copernicus) were emboldened to use small additional epicycles, beyond those discussed here, to fill out details of their models. To a modern reader, it seems bizarre to suppose that the planets are carried on huge invisible wheels rotating in the heavens (the epicycles). But it is important to realize that that was also the usual reaction of anyone but an astronomer for the many centuries during which Ptolemaic astronomy and Aristotelian natural philosophy coexisted. The philosophers treated the epicycles as merely convenient fictions to "save the appearances," and there are repeated calls for an astronomy properly based on the sensible Aristotelian principal of uniform motion around a common center. But nearly all the writings cited to support claims that Ptolemaic astronomy was only instrumental come from philosophers, and none are from important astronomers. The astronomers themselves, conspicuously starting with Ptolemy, write with a good deal of confidence that what they are saying reveals the true structure of the universe.[3] I will try to show why they felt that confidence.

To start, consider how astronomers might come to the bizarre-seeming idea of epicycles. Why might Mars, Jupiter, and Saturn—the planets for which retrogressions are conspicuous—invariably grow brighter as they approach the midpoint of their retrogressions? Or, the same point since the occurrences coincide, why are those planets brightest when in opposition to the Sun? The natural intuition here, based on much familiar experience, would be that the planets are growing brighter because they are coming closer to the Earth as they move through a retrogression. A particularly simple device could then be conjectured to produce not only retrogressions, but retrogressions with the required (by observations) coincidence with opposition, and also the required coincidence of maximum brightness with the center of the retrogression. All this would occur if there was a secondary orbit (an epicycle) that rotated at just the speed needed to match the Sun's motion relative to the Earth.

Figure 7.2 shows the Ptolemaic setup for the Earth, the Sun, and Mars. The planet rides on an epicycle, which in turn rides on the deferent. The deferent circle you see in the diagram is the locus of the center of the epicycle as it is carried around by a disk (or sphere), though it is convenient to wait a bit to introduce the physical setup a Ptolemaic astronomer would have in mind as the basis of diagrams like that of figure 7.2.[4]

At a point on the epicycle away from the Earth, the primary motion of the planet supplied by the deferent is augmented by the secondary motion provided by the epicycle. So as seen from the Earth, the planet would be moving faster than its average speed. At the opposite position, the secondary motion is moving the planet *against* the main rotation. So it would move slower as seen from the Earth, and the effect would be strongest when the planet is closest to the Earth on the epicycle.[5] And other things equal, the bigger the ratio of the radius of the epicycle to the radius of the deferent, the larger the effect of the epicycle would be. Similarly, the faster the rotation relative to the rotation of the deferent, the stronger the effect would be.[6]

But the period of the deferent is fixed by observation of the mean time it takes the planet to return to a given position relative to the fixed stars. The astronomer has no choice about what period to assume. Only one choice will match observations. And the period of the epicycle also is fixed, since retrogressions must coincide with opposition, hence the period of the epicycle must just match the Sun's motion relative to the Earth. So the only parameter that can be adjusted to try to make the size of retrogressions fit observation is the ratio of the radius of the epicycle (r) to the radius of the deferent (R). The best fit that could be achieved determines that ratio. Notice that on the logic so far it is an open question whether anything that would qualify as a good fit could be achieved for any planet. If good fits were possible for all, that would seem quite a gift from heaven.

Imagine an arrow pointing from the Earth to the Sun, and another pointing from the center of the epicycle to the planet (as in figure 7.2). Since retrogressions are observed to always occur at opposition, the required arrangement is that the arrows move exactly together, as described in the legend for figure 7.2. So where would the planet itself be when (in the diagram)[7] the Sun is due north of the Earth? It can only be in one location, namely, due north of the center of the epicycle, as illustrated in figure 7.2. And since this logic applies also to Jupiter and Saturn, the positions of all three planets on their epicycles must be always exactly coordinated.[8]

Suppose a good fit could be achieved for some choice of $r : R$. Then a model dealing with retrogressions would consist of two linked orbits, what I will call the *p-orbit*, which carries the planet's period (mean time to complete a circuit of the zodiac), and an *s-orbit*, which turns to coincide with the Sun's period. For Mars, Jupiter, and Saturn, the p-orbit carries around the center of the s-orbit, which in turn carries around the planet. So far this gives only a qualitative scheme, and it applies only to Mars, Jupiter, and Saturn.

But for Mercury and Venus, similar reasoning yields an inside-out version of the two-orbit scheme just sketched for Mars, Jupiter, and Saturn. The simplest geocentric explanation of why Mars, Jupiter, and Saturn become brighter during retrogressions and exhibit retrogressions only at opposition is that they are travelling on an epicycle.[9] And the simplest explanation of why Mercury and Venus take more time to move ahead of the Sun (relative to the fixed stars) than to fall behind it is also that they are travelling on an epicycle. For then what an observer on the Earth sees as a simple shuttling back and forth (as if the planet were moving along the dashed path between x and y in figure 7.3) actually involves paths of very different lengths—the long upper arc from y around to x versus the short return on the lower arc from x back to y.

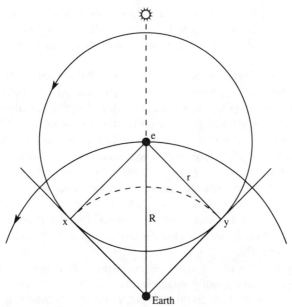

Fig. 7.3. Ptolemaic diagram for Venus. Here (as for Mercury) the s-orbit is the deferent, with a radius always parallel to—in fact, that directly coincides with—a line between the Earth and the Sun. In all the figures, the radius of the larger orbit is R, and the radius of the smaller orbit is r. As explained in the text, from observations an astronomer can calculate what the ratio $r : R$ must be, so that if a value could be defined for either r or R, the other value would also be known, and the dimensions of the entire system would follow. Ptolemaic astronomers did this in one way, Copernican and Tychonic astronomers in an alternative way.

That the ratio of moving-ahead phase to falling-behind is small for Mercury implies that the radius of the epicycle is small compared to the radius of the deferent (so the angle seen from the Earth is small); and the converse for the much larger ratio of moving-ahead to falling-behind phases for Venus.

These Ptolemaic models for Mercury and Venus are inside-out compared to the equivalent models for Mars, Jupiter, and Saturn, but the same elements work for both types of planet. For Mars, Jupiter, and Saturn, the p-orbit is the deferent and the s-orbit is the epicycle; the converse is true for Mercury and Venus. For each planet, the radius of the s-orbit moves parallel to the radius connecting the Earth to the Sun. This yields the parallel positions of the planet relative to the center of its epicycle and the Sun relative to the Earth illustrated in figure 7.2. For Mercury and Venus the same condition holds, but a diagram does not show two parallel vectors, since for these planets the condition takes the trivial form of vectors that lie right on top of each other. The radius of the s-orbit is parallel to the Earth-Sun radius in the trivial sense that it starts at the same point (the Earth) and points in exactly the same direction.

In sum, for every planet, the most striking qualitative features of its motions are captured by a two-orbit model in which one orbit carries a period peculiar to the planet (the p-orbit), and the other (the s-orbit) is tied exactly to the motion of the Sun—but still with the proviso (which of course Ptolemy could satisfy, or we would not be so interested in his models, but which we have not yet considered how to satisfy) that a choice of $r : R$ for each planet can be defined in a way that actually matches the observed behavior of the planets. Once we see how to meet that final condition, we will have not only the periods that must be assigned to p-orbit and s-orbit for each planet but also, as we will later see, a plausible scheme for determining the size and distances of the planets as well.

By looking at a planet—even looking at it with a telescope if you had one—you cannot see how far away it is. If observation of angles were exact enough, you could indirectly see the distance by a parallax calculation from the shifted apparent location relative to the fixed stars as the planet is watched over the course of a night. But there was no chance of doing that within the limits of naked-eye observations.[10] So it is very important that the two-orbit scheme now sketched turns out to provide a basis for calculating distances—and stretching a bit further, even sizes—of the planets, not just angles of observation. It can yield not only effective predictions but an elegant cosmology.

We can use figure 7.3 again to show roughly how to calculate the distance for the particularly simple case of Venus. Approximating the extreme angle between Venus and the Sun as 45 degrees, the span of these extreme positions for an observer at the Earth is a right angle. Since the extreme positions (at x and y) must be tangent to the presumed epicycle, angles Exe and Eye also must be right angles. So the angle at e also must be a right angle. Hence the ratio of the lower arc from x to y to the return upper arc from y to x must be 1 : 3, which indeed fits observations of the times required for the slow and fast phases. The planet indeed behaves as if it were riding on an epicycle, not simply shuttling back and forth along the arc xey. Since the figure formed by the straight lines in the diagram is a square, the ratio of the radius of the epicycle (r) to the radius of the deferent (R) is the ratio of the side of a square to its diagonal, or about 1 : 1.4. Similar procedures define the analogous ratios for the other planets.

But the situation was complicated by variations of the sort described in connection with figure 7.1.[11] For a long time (from the invention of epicycle models about 225 CB.C. to Ptolemy's work around A.D. 150) no one saw how to frame the models in a way that would provide a convincing match to actual retrogressions. The very natural doubt that something as strange as epicycles could be real presumably played a role, discouraging hard work on the possibility that they could be made to yield finely detailed correct results. But once Ptolemy somehow succeeded, then for astronomers—who alone understood how strikingly inferences drawn from epicycle models in fact did work— belief grew that the epicycles were real.[12] They were too pretty to be wholly wrong, and as we will see, they were not wholly wrong. Indeed they were brilliantly right. To this day we continue to see things (from the Earth) in terms of a two-orbit model, but now the s-orbit is the Earth's own orbit around the Sun, not something physically bound to the p-orbit of each planet.

To sum up the difficulty that astronomy before Ptolemy faced: Retrogressions vary in size and shape, as illustrated in figure 7.1, and in their spacing against the background of fixed stars. But in the scheme illustrated by figure 7.2, the retrogressions would be equally spaced (the distance from the midpoint of one retrogression to the next would be fixed), and they would always have the same shape. For nearly four centuries—until Ptolemy was finally able to resolve the difficulties—the epicycle idea remained a mathematically elegant way to give a qualitative account of how planets moved, but could not actually match what happened. Nevertheless, as is commonly the case, once we know how the problem was resolved, it is not very difficult to sketch the main lines of the solution.

Look back at the sample of retrogressions shown in figure 7.1. Sometimes the retrogression is fat and sometimes thin. The span of the loop varies. And the time interval and also the distance along the zodiac between successive loops varies. In the Copernican system the retrogressions are nothing but parallax effects. But suppose there really were epicycles generating the loops. The way the motions appear to an observer on Earth would depend on the angle and distance from which he sees them. You can think of the epicycle as like a phonograph record with a lighted spot on the perimeter, turning while being carried across a dark room so that all we see is the lighted spot. The loops the lighted spot traces out will be contingent on the varying angles and distances from which we see the rotating platter.

The thick versus thin variation (as Kepler was the first to understand) occurs because the planes of the orbits of the planets are tilted slightly relative to the plane of the Earth-Sun orbit. When the Earth is near the intersection of its plane of motion and that of a planet (the node), then the retrogression will be thin: we see the epicycle from on edge. When the Earth is well away from a node, the retrogression will appear fat.

The variation in size of the retrogressions and the variation in distance (against the background of fixed stars) of successive retrogressions can then be understood as due to variations in distance between the Earth and the center of the epicycle. Suppose you watch a planet during times when it is halfway between opposition and conjunction (so it is neither slowed down nor speeded up by the motion of the epicycle but just moving at its mean motion). The speed still turns out to vary, but now in a way that depends on the planet's position in the zodiac.[13] The simplest account of that is to suppose that the deferent moves uniformly but is centered on a point (the eccentric) somewhat removed from the point of observation. Then as viewed from the Earth, a planet in fact moving uniformly will appear to move faster when it is closer to the Earth, and slower when it is more distant. The epicycle will then appear larger when the planet is moving faster (since the center of the epicycle will be closer to the Earth) and smaller when the planet is moving slowly.

In fact, that is the right qualitative variation, and would also account for part of the required variation in spacing of the retrogressions. But quantitatively, the correction for the size of the retrogressions is too strong, and the spacings still do not come out quite right, so that centering the deferent on an eccentric point improves the empirical fit but still does not quite work. The eccentricity can be set to get a good fit with the variation in speed through the zodiac, but this then conflicts with the observed variations in the apparent size of the epicycle. Roughly speaking, the retrogression is now too small to fit observations when it occurs far from the Earth, and too big when close.

Seeing what else might be done was difficult. As Copernicans, we know that what is happening involves both a variation in the distance of the planet from the Sun and also a variation in the Earth-Sun distance. But a pre-Copernican astronomer would see only a variation in the speed of the planet (assumed to be in orbit around the Earth), since the variation in the Earth-Sun distance should have no effect on how a planet's motions are seen from the immovable Earth. Several centuries went by before Ptolemy refined the analysis to the point where he could see a remarkably effective resolution of this conflict through the device that (much later) came to be known as the *equant*.

Things work out very precisely right if the eccentricity that gives the correct variation in speed is divided in half, leaving the Earth

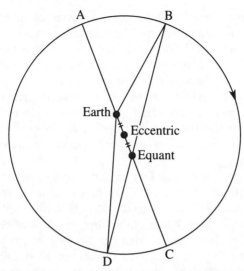

Fig. 7.4. Ptolemy's equant. A, B, C, and D are positions that would successively be occupied by the center of the epicycle carrying the planet as the epicycle is carried around by the deferent. The center of the deferent is *eccentric* to the center of observation (the Earth). The planet's motion appears uniform if viewed from a third point: the *equant*. The eccentric point lies at the midpoint of a line connecting the equant and the Earth. The pairs of sectors marked from the Earth and the equant show the angular change of a point on the deferent an observer would see in a unit of time. A hypothetical observer at the equant would see the movement from A to B and from C to D as taking the same time, though arc AB is obviously larger than arc CD (so the center of the epicycle is actually moving faster from A to B). From the actual center of motion at the eccentric point, the motion would be seen without distortion and as nonuniform. From the Earth, the nonuniformity would be exaggerated: an actual observer would see the movement from A to B as considerably faster than the movement from C to D, since the angular motion from A to B as seen from the Earth is obviously larger but takes the same time to traverse as moving from C to D. As described in the text, motions predicted by this odd-seeming arrangement in fact closely approximate Kepler's laws for the actual motion of planets.

at one endpoint and using the other endpoint as the point from which rotation of the deferent would appear uniform (figure 7.4). This creates a much-discussed problem for physical interpretation, since motion of the epicycle around the actual center of the deferent is no longer uniform. A clue to why this setup works so wonderfully comes from noticing that we now have a model that is virtually indistinguishable from one where the planet travels on a nearly circular ellipse with its central body (the Earth for Ptolemy, of course; the Sun for us) at one focus of the ellipse. Under Newton's theory, or Kepler's empirical fit, the empty focus turns out to be a point from which the motion of the planet would look uniform. By ingenuity and shrewd analysis of observations, Ptolemy had found a convenient-to-calculate functional equivalent of the Kepler ellipse, highly accurate within the limits of naked-eye astronomy.

Copernicus, apparently following earlier Arab work, provided an additional bit of mechanism (further small epicyclets) that allowed the equant motion to be reduced to the interaction of circles uniformly turning on their proper centers. This greatly impressed Tycho and many others who were not themselves Copernican, since it provided what seemed to them a plausible physical interpretation of the equant motion.

But Kepler, from his first publications, returned to Ptolemy's equant and (with the help of Tycho's observations) eventually transformed this Ptolemaic device into the form that was exploited by Newton. The Arab work shows that eliminating the equant did not lead an astronomer to the heliocentric idea (as a review of the previous paragraphs will show, there is no logical connection at all between the two ideas); and Kepler's lack of concern about the equant shows that eliminating the equant was not a necessary or even necessarily an attractive idea to a Copernican.

Yet though there is no direct connection between the two, somehow Copernicus's interest in the equant led him to see the heliocentric possibility. In *Patterns* I gave an account of how Copernicus may have hit on one idea while working on the other. In chapter 9 I add a curious bit of direct evidence for that.[14]

The last piece of machinery we need to consider is only implicit in Ptolemy, but is essential for seeing the intimate connection between Ptolemaic and Copernican astronomy.

A modern reader starts from thinking in Copernican terms, and from that point of view it is obvious why Ptolemaic orbits must be compounded from two interacting orbits. One orbit is the planet's actual motion. The other captures the effect on observations due to the fact that we are observing from a moving platform (the Earth) that is itself in orbit around the Sun. Or in the language I have been using: from a Copernican point of view, it is obvious why observations turn out to fit models that capture the joint effects of a p-orbit and an s-orbit. But if the world is Copernican, the s-orbit (the Sun's apparent orbit around the Earth) and the p-orbit (the planet's orbit around the Sun) are independent. Yet in the Ptolemaic system one orbit carries the other around. This inserts a link between the two orbits that has no physical counterpart. To make things come out right, we need a countermotion that undoes that damage.[15]

Think of the motion to undo this artifactual Ptolemaic linkage, so that the proper rotation of the planet on its epicycle becomes independent of the motion of the deferent, as like the motion of a chair on a Ferris wheel. The epicycle must somehow be arranged to always maintain a fixed orientation. This requires turning backward—for the Ferris wheel, making the chair heavy on the bottom assures that gravity will do the job—at just the rate the main wheel is turning forward.

Or think of the epicycle as like a track, and the planet as like an engine moving along the track. The linkage motion for the epicycle is a backward rotation of the track itself—the Ferris-wheel motion—just equal but opposite to the motion of the deferent. The second component is the planet's own forward motion: just equal to the Sun's period around the Earth if the s-orbit is the epicycle; just equal to the planet's own period if the p-orbit is the epicycle. Figure 7.5a shows the conveniently simple case where the center of the epicycle is due north of the center of the deferent, and the planet (Mars) is due north of the center of the epicycle. Since the p-orbit is the deferent for Mars and has a period of two years, three months later R has moved 45 degrees (one-eighth of a complete cycle), as shown in figure 7.5b. But the Ferris-wheel (backward) motion of the epicycle keeps the open circle, marking the point at which Mars was in figure 7.5a, still due north of the center of the epicycle. The (forward) motion of the epicycle—which for Mars is the s-orbit, with a period of one year—moves the planet one-fourth of the way around the epicycle over the three months. The net effect, which indeed matches observations of Mars, is shown in figure 7.5b.

Yet another way to envision the linkage is to imagine the epicycle as connected to the deferent by a frictionless bearing. If the epicycle were not turning at all, a creature at the bearing would see the epicycle rotating backward, though actually it is just sitting in its inertial position while the bearing slides by. But the epicycle has a motion of its own (for Mars, its period of two years), so the creature at the center would see the joint effect of the forward rotation, partly offset by the apparent backward rotation of the linkage motion.

The same sort of arrangement shown in figure 7.5 for Mars works also for Jupiter and Saturn. For Mercury and Venus, since the s-orbit is the deferent and the p-orbit is the epicycle, a converse arrangement holds. The linkage motion then provides a Ferris-wheel adjustment to the p-orbit to offset the motion of the s-orbit, instead of an adjustment to the s-orbit to offset the motion of the p-orbit.

The same result can be obtained, though at a cost in other respects, without making the linkage explicit. Within Ptolemaic astronomy, in fact, no linkage motion is invoked. The Ptolemaic scheme for

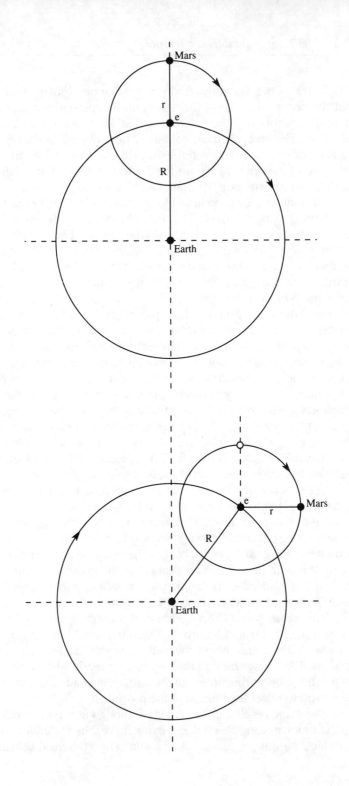

Mercury and Venus simply takes the period of the epicycles for these planets as the time to return to a given position relative to the Sun as seen from the Earth. The apparent periods of Mercury and Venus were not interpreted as their true periods offset by a linkage motion, but just as their true periods. Similarly the epicycles of Mars, Jupiter, and Saturn were not taken to have periods equal to that of the Sun, each with the linkage adjustment set by its p-orbit. They simply had a natural motion, different for each planet, such that the position of each on its epicycle always coincided exactly with the Sun's position relative to the Earth.

Logically, Ptolemy could easily (without making anything we would see as a striking discovery) have used the linkage motion sketched here. That would have had a certain advantage in terms of enhancing the coherence of his system, as will be shown later. But Ptolemy and his successors over many centuries never mention this interpretation. We will soon want to consider why. But for us, always aware of the Copernican possibility, taking explicit notice of the linkage motion is essential. It makes better intuitive sense of the Ptolemaic setup, and it helps us see why the Ptolemaic system can be transformed into an observationally equivalent Copernican or Tychonic system.

7.4

Working through the detailed Ptolemaic geometry is complicated enough that mastering the system in the way needed to work with it—rather than (as here) seeing only what the basic logic is—would require extensive study. The number of people alive at any time who had mastered it must always have been very small. Logically, such a scheme should work *qualitatively:* it has the right sorts of properties, as we have been seeing. But there is no reason at all to suppose that parameters can be fit that will closely match the way the planets move.

Fig. 7.5. The Ferris-wheel linkage illustrated for Mars. The p-orbit is the deferent, which takes two years for a complete rotation, and the s-orbit is the epicycle, so it takes (of course) one year to rotate. In *a* the center of the epicycle happens to be due north of the Earth, and Mars is due north of the center of its epicyle (so you know from the discussion in connection with figure 7.2 that the Sun must also lie due north of the Earth even though it is not shown in the figure). Three months later (one-quarter of a year; *b*), the p-orbit has moved 45 degrees (one-eighth of a cycle). The center of the epicycle is now northeast instead of north of the center of the deferent. The combined effect of the Ferris-wheel linkage (turning the epicycle backward to just offset the motion of the p-orbit) and the s-orbit's annual motion (forward ¼ of a cycle over three months) puts Mars now due east of the center of the epicycle, as it must be to match observations.

That a close fit could actually be found provided a powerful inducement to belief that a system so elegant *and* so empirically effective must be essentially right.

Within the limits of naked-eye observation, Ptolemy could account for all the major variations in observations. The lasting power of Ptolemaic astronomy until the Copernican idea came along is therefore not surprising. The system works, and relative to the complexity of the observations it accounts for, it is remarkably simple. Substantial errors appeared only over long periods of time that allowed small discrepancies to accumulate, implying that it was almost, but not quite, perfect. That is why, for a Ptolemaic astronomer, the deferent-plus-epicycle apparatus would become irresistibly real, as for a contemporary physicist the experience of observations and manipulations of oscilloscope images and the like remove any tinge of subjective doubt that electrons and so on are real, whatever philosophers may think about the matter.

But now consider some details. Only the ratios of s-orbit to p-orbit have been fixed by the argument so far. Figure 7.2 shows the epicycle for Mars lying outside the orbit of the Sun, and figure 7.3 shows the epicycle for Venus lying inside the Sun. That indeed is the Ptolemaic arrangement. But models equally consistent with observations and the $r : R$ ratios they imply might set Mars between the Moon and the Sun, or let the center of the epicycle for Venus coincide with the position of the Sun itself (so the orbit of Venus would then be heliocentric), or even put Venus beyond the fixed stars.

However, suppose that some principle for scaling was in fact available, so that for each planet the logic of the system somehow determines the size of either the p- or s-orbit; then since we know the ratio $r : R$, the size of the other orbit is also determined. We would then have distances (for each planet), no longer just a ratio of distances. The various Ptolemaic models for individual planets could be scaled to this common principle to yield a system of the world, of known dimensions. In particular an astronomer could seek to set the scaling such that the individual planets fit together into some cosmologically interesting scheme—where, for example, neat connections and physical interpretations of the formal models come to hand. And in fact not one but three such schemes play a role in the history of astronomy (Ptolemaic, Copernican, and Tychonic). We will also have occasion to consider a fourth (what I will call the *inverted* Ptolemaic system), which is logically and aesthetically interesting but curiously absent from the record.

We start with the solution Ptolemy himself proposed, which turned out to survive essentially unchallenged for 1,400 years.

Suppose that (1) there is no unnecessary space in the universe, and (2) the spaces required for the various heavenly bodies do not overlap. A system of the world can then be constructed as a set of snugly nested spheres, where the planets are arranged such that the spheres (or disks)[16] that account for the observable motions (the spheres themselves being invisible) neither collide nor leave unnecessary space.

For each successive planet, the outer limit of the space required for the previous planet would uniquely determine the inner radius of its sphere. Since the ratio of epicycle to deferent ($r:R$) has been determined from observations,[17] the outer boundary of the sphere for a planet is then also defined. The sphere has to be thick enough to contain the epicycle, given the required ratio $r:R$ and the minimum value of the inner radius of the sphere.

So far this leaves the order of the planets undetermined. Ptolemy provided plausible reasoning to settle that. The Moon, alone among the heavenly bodies, is close enough for naked-eye instruments to observe its parallax, and hence to calculate its distance. This varies over the course of a lunar cycle, and in other ways the motion of the Moon is more complicated than that of any planet. We would explain that motion in terms of the joint gravitational effects of the Earth and Sun. In Aristotelian terms, followed here by Ptolemy, the complicated motion of the Moon (compared to the planets) reflected its adjacency to the unstable terrestrial realm. But since the Moon's maximum distance could be computed, a Ptolemaic astronomer knew where the inner edge of the nearest planet must come.

The nearest planet, in turn, was most plausibly Mercury, whose motions are more complicated than those of the remaining planets, revealing that it too is somewhat subject to the influence of the Earth.[18] The outer distance of the sphere of Mercury is then unambiguously determined by the size of the epicycle required to fit its motions to observations, given that we know the inner radius. Since the successive spheres are *snugly nested,* the inner radius for the next planet is Mercury's outer radius. If we make the further plausible assumption that the slower the deferent of a planet turns, the more distant it must be, the rest of the heavenly bodies can be put into place.[19]

Mars, Jupiter, and Saturn must be the three most remote planets, in that order. This leaves an ambiguity only for the Sun and Venus, which (along with Mercury, which has already been placed just beyond the Moon) have the same period. Snug-nesting demands that the ambiguity be resolved by putting Venus between the Sun and Mercury, since a huge gap would be left if the Sun were put inside of Venus.[20] The Ptolemaic world is built out from the center (the Earth) like a spherical layer cake, determined by observations of cycles and retrogressions, plus some plausible reasoning about the order of the planets, plus the nesting principle.

The appeal of this scheme was confirmed by its neat symmetry. The Sun is in the midst of the planets, dividing them into an inner set (Mercury and Venus, both closer to the Earth than the Sun is) and an outer set (Mars, Jupiter, and Saturn, all lying beyond the Sun). The motion of the strategically placed Sun governs the deferents of the inner planets and the epicycles of the outer planets. So we have a reason why the deferents of the inner planets and the epicycles of the outer planets move to coincide with the Sun's motion. Hence the two classes of planets determined from fitting epicyclic models are the same two sets that the Sun's location divides. Each set—the inner planets, the outer planets—is governed in a characteristic way by the motion of the Sun around the Earth (*Patterns* [11.8]).

Another symmetry appears when you notice that the size of the epicycles grows smaller as you move away from the Sun in either direction. Even more striking, an apparently independent computation showed that the space required for the two inner planets just matches the space available between the Moon and the Sun. We know today that the startling neatness of this last fit is an illusion. But the overall setup was so convincing that the crucial fit of Mercury and Venus to the separately calculated space available between the Moon and Sun was never questioned until Ptolemaic astronomy was already dead from other causes. The fit became, as Neugebauer (1982, 111–12) remarks, the unquestioned foundation of the physical astronomy challenged by Copernicus. See note 20.

A complete Ptolemaic world emerges, where distances and (from the distances and apparent angular diameters) the sizes of all the major bodies are unambiguously determined, neatly tied together by what could be seen by careful observation of the heavens and a few plausible assumptions. All of this was strikingly validated by the neatness with which everything fit together, and by the astonishing (by the standards of the time) accuracy of its predictions.[21]

The basic models for the planets themselves are as simple as one could hope, with just one circle to represent the solar motion that affects each planet, and one circle to represent the planet's own motion. The equant/eccentric scheme sketched earlier nicely accounts for the complications introduced by the planets' anomalies in progress through the zodiac. These mathematical models then lend themselves to the neat physical interpretation illustrated by figure 7.6, where the complexity introduced by eccentricity of the deferents itself turns out to have a ready physical interpretation (discussed in the legend). The in-

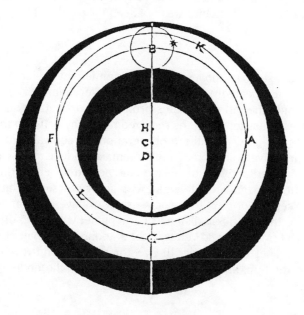

THEORICA ORBIVM ET
centrorum trium superio-
rum,& Veneris.

ß

Fig. 7.6. A Ptolemaic sphere for Venus. The drawing is from a fifteenth-century text, but the conception dates back to Ptolemy. Each component of the sphere provides a component of the motion: the epicycle provides the p-orbit, the deferent provides the s-orbit, and the dark shells that fill out the sphere, making it properly centered on the Earth even though the deferent is eccentric to the Earth, provide the daily rotation. Comparing this diagram to figure 7.4, the Earth is at D, the equant point is at H, and the eccentric point is at C. The center of the epicycle is at B, with the planet's location indicated by the star on the epicycle. The circle traced out by B as the epicycle is carried round inside the rotating deferent is BFGA. FKAL is the equant circle, centered on H. Reprinted with permission from *Vistas in Astronomy* 17.

dividual planetary models can be scaled into a scheme in which everything fits neatly together, in spheres properly centered on the Earth, with no crowding and no wasted space.

The system is in a sense *perfect.* At its demise, as at its birth, the Ptolemaic arrangement was as accurate as it was realistically possible to be within the limits of naked-eye astronomy.[22] But as we will see next, geometrical moves that are not at all difficult compared to other details of Ptolemaic geometry allow many other observationally equivalent possibilities, two of which (the Copernican and Tychonic), once on the stage, proved capable of winning leading astronomers away from Ptolemy even before there was serious empirical evidence against him.[23] Two of the moves date back to the early history of the epicyclic idea, long before Ptolemy. The third was not seen for another 1,800 years.

7.5

The first of the moves is so very simple that we have implicitly already used it repeatedly. Since naked-eye observations of a planet can determine only the ratio between s-orbit and p-orbit, then among possible choices consistent with the required ratio (say between the pair of possibilities in figure 7.7) all are observationally equivalent. So an immediately intuitive point is that any Ptolemaic model can be *scaled*. We have already used one way of scaling (the nested-spheres principle) in working out the standard Ptolemaic cosmology. A second interesting possibility sets all of the Ptolemaic s-orbits to have a common radius, just equal to the Earth-Sun distance. For a planet with the p-orbit riding on the s-orbit (Mercury and Venus in the standard Ptolemaic setup), this makes the planet heliocentric: the vector of an s-orbit always points directly to the Sun, and this scaling move makes the endpoint of the

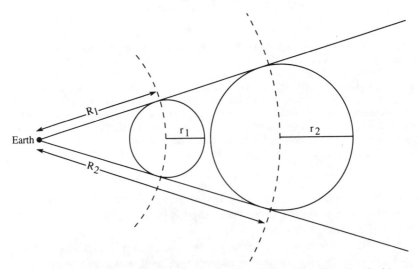

Fig. 7.7. The scaling move. Observations determine only the ratio $r:R$ as sketched in section 7.2. Any value for R will work, so long as r is chosen to make the epicycle just fit the observed angle. Alternatively, any value of r will work, so long as R is appropriately chosen. The two arrangements shown would be observationally identical, since $r_1:r_2 = R_1:R_2$.

vector coincide with the Sun. That is the situation shown for Venus in figure 7.10, which can be compared with figure 7.3 showing the usual Ptolemaic arrangement. It seems that the s-orbit has disappeared, since it coincides with the Sun's orbit. A person with Copernican habits of mind sees the planet in the position it ought to be, namely, heliocentric. For the remaining planets, where the s-orbit rides as epicycle, scaling gives the epicycle a radius just equal to the Earth-Sun radius. The snug-nesting characteristic of Ptolemaic cosmology is then lost. But in a moment, we will see how that sets the stage for a startling transformation.

Inversion is not so obvious as scaling, so that a little work with pencil and paper (sketching the diagrams) will help in seeing just what is going on. The usual way of arranging the Ptolemaic models is to have the smaller orbit ride on the larger. For what I will call the "standard" arrangement, where the smaller orbit rides on the bigger, the epicycle is the p-orbit for the inner planets (Mercury and Venus), and the epicycle is the s-orbit for the outer planets (Mars, Jupiter, and Saturn).

Since the Ferris-wheel linkage (figure 7.5) keeps the turning of the orbits independent, it does not in fact make any observational difference which orbit is set as deferent and which as epicycle. The Ptolemaic models can be turned inside out, or *inverted,* with no effect on the fit to observations, provided only that the linkage is correctly set. This was understood almost as soon as the epicyclic idea was invented, and so predates the Ptolemaic system itself by nearly four centuries.

The solid lines in figure 7.8 show the positions of R and r in figure 7.5b, with R pointing to the center of the epicycle and r pointing from the center of the epicycle to the planet. Since we are going to consider inversion of the standard arrangement, understand R to be the radius of the larger circle (whether in the deferent position or riding as the epicycle), and consider r to be the radius of the smaller circle.[24] In the standard (small-on-big) configuration, R_{st} in figure 7.8 connects the center of the deferent with the center of the epicycle. And at the end of r_{st} we find the planet (here, Mars).

Now suppose we turn the model inside out, so that the big circle rides on the little circle. This runs contrary to habits of mind well entrenched by experience in the world, which make us familiar with little things that are carried around by big things, not the converse. With a little practice a reader will be able to envisage how the standard (small-on-big) configuration produces retrogressions. But even with a good deal of practice it remains uncomfortable to do that for the inverted (big-on-small) configuration. Nevertheless, it is not hard to see that the logic of the situation guarantees that in fact the motions generated by

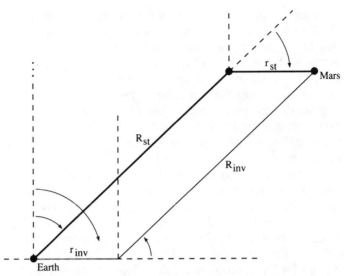

Fig. 7.8. The invert move. R rides on r, the reverse of the standard Ptolemaic arrangement. As described in the text, this figure shows standard and inverted schematics for the situation in figure 7.5b. From observation astronomers had found that the larger orbit for all the planets has the slower intrinsic period. The inverted model then seems doubly odd. It violates our intuitions about little things riding on big, not the reverse, and the second orbit appears to turn backward relative to the common motion of all the primary orbits and of the Sun around the Earth. In the standard arrangement, R_{st} moves clockwise (as this figure is drawn), and r_{st} also moves clockwise even from the position it would have if (following Ptolemy) the linkage was left implicit. But R_{inv} is tilted back from the dashed line that shows the continuation of r_{inv}. It seems to move counterclockwise. As explained in the text, the planet nevertheless ends up in exactly the same position whether we follow the standard arrangement, or invert that arrangement. Since the slower orbit rides on the faster when the model is inverted, the linkage counterrotation, which only slows the forward motion in the standard setup, now is so large that the apparent motion reverses.

the standard and inverted schemes are identical. The lighter lines in figure 7.8 show the situation for the inverse setup.

Imagine the standard model as a clock with a single hand (R) and a blank face (so you can tell where the hand is pointing only if you know that the twelve-o'clock position is at the top). At the end of that hand imagine a watch, also with a single hand (r) and a blank face. Suppose that, as in figure 7.5a, both R and r point to twelve. Consider what happens if there were no Ferris-wheel linkage. When R on the clock turns 45 degrees (reaching the position shown in figure 7.5b), r in the watch would still point to twelve, but the whole watch would have turned so that the hand points northeast instead of north. While R (with a period of two years for Mars) is turning 45 degrees, r (with its period of one year) is turning 90 degrees. So without the linkage motion the total effect would be that r (the hand in the watch) would now point southeast. If we constructed a two-orbit model that works that way, it would radically fail to match observations. So we need to insert the linkage motion, or something functionally equivalent, as described in connection with figure 7.5. The linkage turns the watch backward 45 degrees as R moves forward that amount, leaving twelve on the watch always pointing north, whatever the location of the watch. Now the hand (having moved 90 degrees) points east, as shown in figure 7.5b, not southeast, which is what is required to match observations.

Next think of the inverted scheme. You must now imagine a very sturdy watch that can carry the clock at the end of its hand. The watch hand (which is still the s-orbit, and thus turns 90 degrees in the three-month interval from figure 7.5a to figure 7.5b) moves to point east as before, but since it is now at the center, not carried around by another motion, it no longer needs a Ferris-wheel counterrotation to keep things right. This radius is shown as r_{inv} in figure 7.8. It carries the clock around so that the clock's hand (aside from its own motion and the linkage motion) would also point east. So we need the counterrotation to turn the clock back so that its twelve position remains properly oriented to the north. The hand's own motion then moves R_{inv} 45 degrees forward. But if the linkage is left implicit, R_{inv} seems to move backward, as shown in figure 7.8. In either treatment, the motion of the planet as seen from the center is exactly the same.

Figure 7.9 shows a physical representation (the analogue of figure 7.6) of how an inverted model would work for a case where the radius of the larger circle (R) is greater than the diameter of the smaller circle $(2r)$. Figure 8.1 shows the opposite case where $2r > R$.[25] In either case, the smaller orbit is now the mathematical locus of the motion of the center of the larger circle. The shaded disk is the deferent, carrying the larger circle as an epicycle.

Comparing the inverted physical model of figure 7.9 with the standard model of figure 7.6, the epicycle is the rim of a wheel that encircles the center, rather than a rim or disk that is contained within one sector of the deferent. For $2r < R$ (as in figure 7.9), the locus of the center of the epicycle as it is carried around lies wholly within the inner boundary of the shaded disk. For $2r > R$, the locus crosses into the shaded area (as in figure 8.1), and we get the apparent intersection of orbits that so troubled Tycho. But either way, the inversion does not change anything observable. Further, a nested-spheres inverted model

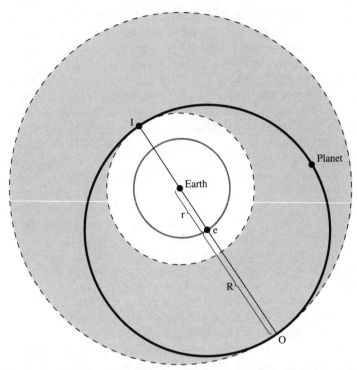

Fig. 7.9. An inverted Ptolemaic physical model with $2r < R$, so the radius of the larger circle (R) is bigger than the diameter ($2r$) of the smaller circle. The small circle (with radius r) is lightly indicated as a reminder that it is only a mathematical locus, as the R orbit is in the standard model. It is not something that could be bumped into, any more than the equator or a center of gravity could be bumped into. The sphere of the planet is bounded by the inner and outer dashed circles defining the shaded disk. The radius of the inner boundary $R - r$, and the radius of the outer boundary is $R + r$, exactly as it would be in the standard configuration. A representation fully parallel to the fifteenth-century figure 7.6 would also show the equant and the eccentricity of the deferent. The region lying between the dashed circles passing through I (inner boundary) and the dashed circle passing through O (outer boundary) would rotate, carrying within it the epicycle (bold ring passing though both I and O), which itself is rotating, carrying the planet.

like figure 7.9 is the same size as the standard model, as explained in the legend. So if figures 7.6 and 7.9 were standard and inverted setups for the same Ptolemaic planet, and drawn to the same scale, then the complete spheres for the two versions could be made to coincide exactly.

We now have three basic cases for geocentric models: the standard Ptolemaic models, where for each planet we have the usual small-on-big configuration, giving the physical setup of figure 7.6; and two classes of inverted model: for $2r < R$, the setup of figure 7.9 with the smaller orbit wholly inside the large; and for $2r > R$, the setup shown as figure 8.1, where the smaller and larger orbits intersect. But to repeat that important point, the inverted smaller orbit is only the mathematical locus of the center of the larger orbit, which itself is carried around entirely inside the deferent. The smaller orbit itself is not a physical thing that could collide with something.

What is the invert move good for? In fact it lets us specify two cosmologically interesting systems, both discussed in chapter 8: the Tychonic and a new inverted Ptolemaic system. Thinking through these possibilities will sharpen your insight into the transformation of the Ptolemaic system into the Copernican. But the invert move cannot reach the Copernican system, nor is it essential for it. What is essential is a third move, *sliding*, which in contrast to *scaling* and *inversion* seems never to have been noticed before Copernicus.

Start with the simplest case, considering only the Sun and Earth (so ignore for the moment the heliocentric orbit also shown in figure 7.10). Take any arbitrary point (S_1) on the apparent path of the Sun around the Earth. If we slide the Sun over to the position of the Earth, as indicated by the solid arrow in figure 7.10, what needs to be done to leave all observations unchanged? If *only* the Sun and Earth are involved, then all that need be done is to move the Earth the same distance in the same direction (the dotted arrow). Similarly, if we consider S_2, the slide again puts the Sun at the center, and the Earth at E_2', and similarly for S_3 and its corresponding point for the Earth at E_3'. Since the same move can be made for every point on the Sun's apparent orbit around the Earth, we get an alternative model in which, instead of the Sun moving on an orbit around the Earth, the Earth moves on the Sun's orbit around the Sun; and the Sun replaces the Earth at the center. The *slide* move, therefore, turns the Earth-Sun model inside out. If there were nothing else in the world to be observed, there would be no clue (from simple observations) which way things really are. The angle at which the Sun is seen from the Earth would remain the same, as would the apparent diameter of the Sun, and the direction in which the Sun is observed to move.

7.6

But now consider what must happen to maintain observational equivalence with respect to the planets if we slide the Sun to the center (pushing the Earth into the Sun's Ptolemaic orbit, as just described). As mentioned at the beginning of section 7.5, the interesting cases here are

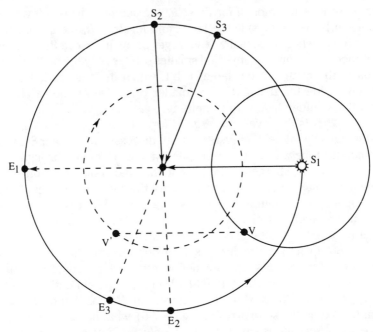

Fig. 7.10. The slide move when p-orbit rides on s-orbit (Mercury and Venus). Scaling the s-orbit to the Earth-Sun distance makes the orbit heliocentric, since the center of the p-orbit here always lies on a line with the Sun. The Sun slides to the center, and all other bodies slide so that their position as observed from the Earth remains unchanged. This requires that each body slide along a line parallel to the Earth-Sun radius, for a distance equal to the Earth-Sun distance. Given the prior scaling move, the slide move matches every point on the Sun's orbit with a mirror-image point of the heliocentric orbit of the shifted Earth. The corresponding slide for points on the planetary orbit shifts the orbit from a Tychonic to a Copernican position, as illustrated by the slide of V to V'.

where the slide move follows scaling of all s-orbits to match the Earth-Sun distance. So (for the standard Ptolemaic setup) Venus and Mercury are then heliocentric, and Mars, Jupiter, and Saturn have identical epicycles, each with a period that matches that of the Sun and with a radius that matches the Earth-Sun radius.

    For the fixed stars, no compensating shift is required when the Sun slides to the center, pushing the Earth into the Sun's former orbit. They are too distant for this to yield any effect observable by naked-eye astronomy. But each planet must slide a distance identical to that of the Sun, parallel to that move, so that the angle from which the planet is seen by an observer on the Earth will be preserved. For a planet that is in the heliocentric position we reached by scaling Mercury and Venus, sliding moves the orbit of the planet along with the Sun, as shown in figure 7.10. But inverting the outer planets puts them also into heliocentric orbits (their p-orbits now ride on their s-orbits), and as with Venus and Mercury,[26] scaling makes the s-orbit coincide with the Sun's orbit around the Earth. So scaling the inner planets together with inverting + scaling the outer planets gives us what we now know as the Tychonic system. Adding the slide move to that gives us the Copernican system. The planetary orbit falls inside the orbit of the now heliocentric Earth if the planet is an inner planet in the Ptolemaic setup, and outside if it is a Ptolemaic outer planet. The linkage motion of the epicycle assures that the planet's orientation (how its eccentricity is arranged in space) does not depend on where the planets are when we execute the slide.

What if the slide move is made after scaling but without inverting the outer planets? The result is identical, as can be seen in figure 7.11. The planet must slide with the Sun, the same distance and parallel

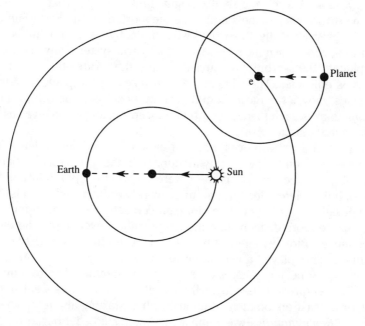

Fig. 7.11. The slide move when s-orbit rides on p-orbit (Mars, Jupiter, and Saturn). Since the s-orbit here is the epicycle, the radius is always parallel to the Earth-Sun line. With the scaling move, consequently, a shift parallel to and equal in distance to the slide of the Sun to the center just slides the planet off its epicycle onto the now heliocentric deferent.

to the shift of the Sun. But from scaling, the radii of the s-orbits are all just equal to the Earth-Sun distance. The Ptolemaic setup guarantees that the radius of an s-orbit—which is the epicycle for these planets—will always lie parallel to the Earth-Sun line [7.2]. So the outer planet slides off its epicycle onto what had been its geocentric deferent, which has now become a heliocentric orbit!

In sum, scaling the s-orbits to the Earth-Sun distance sets the stage for transforming the Ptolemaic system into an observationally identical Copernican or Tychonic system. The slide move takes the Sun from its geocentric orbit and makes a mirror image of that very orbit into a heliocentric orbit of the Earth, carrying the orbits of any heliocentric planets with it. At the same time, any planet in the configuration of p-orbit on s-orbit (the standard configuration for Ptolemaic outer planets) will slide off its epicycle, with its deferent becoming a heliocentric orbit. So whether particular planets are taken in standard or inverted form, with scaling and sliding we get the Copernican setup.

As an exercise, a reader might think through what happens if we invert the *inner* planets and scale the s-orbits to the Earth-Sun distance. Observationally, this would work perfectly well, though it is very hard to see in your mind's eye. We get a hybrid system in which for all the planets the s-orbits are carried around the Earth riding on the p-orbits as epicycles. The slide move then takes all the planets off their epicycles, moves the Earth from its central position to an orbit around the Sun, makes each planet's p-orbit heliocentric, and again gives us the Copernican system.

Of the three moves, only the first was used by Ptolemy. The inversion possibility is explicitly mentioned in the *Almagest* (so there is no doubt that ancient astronomers were aware of the possibility of this move), but Ptolemy does not actually work with inverted models. The slide possibility went unnoticed for eighteen centuries, though logically it seems no more difficult than the long-familiar invert move. So of the three moves, Ptolemy knew of two but only used the simplest, the scaling move. That played a very large role in the story, since it produced the particular nested-spheres cosmology that dominated astronomy for so many centuries. Other logically available cosmologies went unnoticed, or at least uncommented upon. Further, starting about forty years after *De Revolutionibus* was published, Ptolemaic belief began to fade, two decades before the telescope finally made available striking observational evidence against it.[27] A few years later, except where the church was able to enforce its dictum, Tychonic astronomy was also fading, again before there was observational evidence against it.

Overall, with the tutorial material of this chapter in hand, we are left with a considerable number of cognitive questions, of which two are particularly striking: Why would one particular system have taken hold so firmly as to remain almost unchallenged for fourteen centuries? How could belief in that system then collapse while the system was still as observationally perfect as it could be?

Eight

A "New" Ptolemaic System

A habit of mind that makes it difficult to see some things must make it hard to avoid seeing other things, just as a physical habit that makes it difficult to make some moves must make it easy to make and hard to avoid other moves. Choices that seem cognitively hard to avoid are not necessarily the best choices, nor are choices that seem inevitable always actually inevitable (so that no social pressure or individual bias whatever could have deflected them). The point is only that we can identify judgments tied to well-entrenched habits of mind, which would be difficult to avoid. As well-entrenched habits of mind are ordinarily effective in the world, judgments that are difficult to avoid in fact are ordinarily not avoided. We encounter a peculiar but instructive example of these effects when we explore some consequences of the "moves" apparatus we have in hand from chapter 7.

The "new" Ptolemaic system that is the focus of this chapter concerns an inversion of the standard Ptolemaic system, which was *logically* readily available from Ptolemy's time onward, but which does not appear at all in work that has come down to us. We will see that in fact this inversion yields a cosmology that has interesting advantages over the standard Ptolemaic system and also over the Tychonic system that attracted conservative astronomers for several decades after belief in Ptolemy faded.

We are given a novel and particularly sharp form of the question of why the standard Ptolemaic scheme faced no serious challenge for 1,400 years. The totally neglected alternative Ptolemaic system I will sketch logically looks much too interesting to ignore. The Earth remains the stable center of the system. The nested-spheres sense of how the world is built, which on the barrier account is what for so many centuries blocked the heliocentric possibility, is also unchallenged. Hence there is a puzzle about why this system could have been so totally overlooked. On the barrier view, we expect to be able to identify some well-entrenched habit of mind that could account for that. It turns out that the salient candidate is one that illustrates the important point that barrier habits of mind are not *necessarily* things that would come to look wrong if only we could momentarily escape them.

8.2

In the introduction to *Astronomia Nova* (1609), Kepler abruptly dismisses Ptolemy as "exploded" with a passing reference to his reliance on *coincidence* to account for why each planet's motion contains a component that exactly tracks the motion of the Sun. Kepler gives a detailed explanation of his Copernican preferences only against the Tychonic alternative. Since Galileo is equally dismissive of Ptolemy in his 1609 *Starry Messenger,* we have an indication of the prevailing state of belief among astronomers. By 1609, though we are still a couple of years away from the discovery of the phases of Venus, expert opinion seems to have already dismissed Ptolemy.[1]

The taken-for-granted confidence with which Kepler treats a mere mention of the coincidence argument as sufficient to dismiss Ptolemy suggests that this argument was both familiar and important to his readers.[2] The coincidence argument turns on the point that, if planetary orbits are heliocentric (which holds for both Copernicus and Tycho), we get an immediate and necessary link between the apparent motion of the Sun and the motions of the planets. But the planet-by-planet models of Ptolemy just happen to move as if they were somehow exactly linked to the Sun, though there is no suggestion of why such a link should exist.[3]

The "new" Ptolemaic system I will sketch (what I call the *inverted* system) is reached by almost the same steps as we have already used in converting the standard Ptolemaic system into the Tychonic system [7.5]. We can get Tycho's system by inverting the Ptolemaic models of the outer planets *plus* scaling all s-orbits to the Earth-Sun distance. But suppose that we inverted the outer planets, leaving the scaling principle the same as in the standard Ptolemaic system. We would have a system in which for all the planets (not just the inner planets) the p-orbit rides on the s-orbit. In this respect the system is exactly the same as the Tychonic. But instead of setting all the s-orbits to the Earth-Sun distance (the Tychonic scheme), the s-orbits are left spread out on the traditional Ptolemaic nested-spheres logic.

   An example of the resulting inverted sphere for an outer planet was shown in figure 7.9. Figure 8.1 is a slight variant of figure 7.9, covering the case where an apparent intersection of orbits results, as indeed occurs for Mars. As explained earlier [7.5], the space required

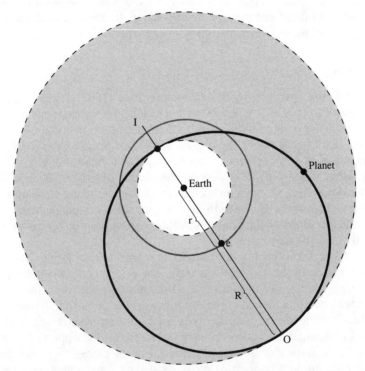

Fig. 8-1. Inverted Ptolemaic model when $2r > R$. The locus of the center of the inverted epicycle here crosses into the deferent disk, in contrast to the situation in figure 7.9, where $2r < R$. But as stressed in the text and notes, the overlap cannot lead to a collision, since the locus is not a physical object but only the path of the center of the epicycle, e. Exactly as in the setup of figure 7.9, the shaded region rotates in place, carrying the epicycle, which itself is rotating in place, carrying the planet. In the Tychonic setup, the locus of the center of the epicycle would also be the orbit of the Sun. But it would again be only a mathematical locus (not something that could collide with the deferent). The entire shaded area would rotate annually, carrying both the Sun and the epicycle around together.

for a planet for the inverted model remains exactly what is required for the standard Ptolemaic model. As discussed below, as a physical problem the intersection of orbits is illusory. So the distances and sizes of the planetary spheres implied by this cosmology are consistent with Ptolemaic reasoning and turn out to be just those of the standard Ptolemaic setup. Similarly, since what determines distances in any of these schemes is the scaling rule, the Tychonic distances are just the same as the Copernican, since the scaling is the same.[4]

If we look at a diagram for a single outer planet, it is almost the same for both the inverted and the Tychonic setups. In both, for every planet, the Sun lies on a line between the center of the epicycle and the geocentric Earth. The difference between the Tychonic and inverted Ptolemaic setups is that in the Tychonic system the s-orbit is scaled to the Earth-Sun distance, so that the Sun coincides with the center of the epicycle: it lies at the e point of the line Ee in figure 8.1. In the inverted system the Sun is on the same line, but just beyond the sphere of Venus and just inside the sphere of Mars, as in the standard Ptolemaic system. So the p-orbits of Tychonic planets are all centered on the Sun, but in the inverted Ptolemaic system, each has its own center—inside the orbit of the Sun for Venus and Mercury, and beyond the Sun for Mars, Jupiter, and Saturn.

Within the limits of naked-eye observation, of course, all the systems (Ptolemaic, inverted Ptolemaic, Copernican, Tychonic) are observationally identical, as shown by the moves analysis in chapter 7. But since the orbits of either the standard or inverted Ptolemaic systems are scaled by spreading out models like that of figure 8.1 so that there is no overlap, we get systems about 50% larger than the Tychonic (or Copernican) alternative.

8.3

Although its dimensions are identical, the inverted system is cosmologically distinct from the standard Ptolemaic system in interesting ways. The crucial aspect of the inverted Ptolemaic system is that in this system, as in the Tychonic but in contrast to the standard Ptolemaic system, a single s-orbit effectively governs the Sun and all the planets together. For either system, the simplest mechanical implementation is if the region of the heavens between the Moon and the fixed stars had a single motion, which would give all bodies within that region (the Sun and the p-orbits of the five nonterrestrial planets) a common annual motion. A Ptolemaic cosmology for the inverted system, but also a physical implementation of the Tychonic system, would suppose that, in addition to the east-to-west diurnal rotation that affects the whole extra-terrestrial universe, there is a contrary annual rotation that affects all bodies between the fixed stars and the Moon. This single motion carries all the planetary orbits, along with the Sun, on an annual circuit. The interaction between that common annual motion and the rotation of the p-orbits of the individual planets then yields the looping paths (retrogressions) which observations demand.

The characterization just given is not an account of how the Tychonic system has been conceived to work. But as I have stressed elsewhere (so I won't go through the technical details again here) that is not because the Tychonic system has ordinarily been conceived to work in some other way. Rather, the question of just how the Tychonic would work has almost never been discussed at all. Tycho himself never spelled out the details, as Galileo was pleased to point out, nor did anyone else.[5] But if you wanted to give a physical account of how the Tychonic system would work, the description of the previous paragraph would apply.

An unexpected consequence of thinking about how the Tychonic system might actually work is that we can see that Tycho was wrong in supposing that his system is incompatible with the traditional conception of the heavens as built up out of solid spheres. As indicated by the preceding discussion, it is straightforward to provide an arrangement in which the supposed collision of spheres does not occur.[6] Further, even if you mistakenly suppose the contrary, that is still no serious argument for Tycho against Ptolemy, since whatever mechanization a Tychonic astronomer might propose for his system would work equally well for the inverted system and could be adapted to work about equally well for the standard Ptolemaic system.[7] Tycho himself never explicitly claimed otherwise.

A *single* motion to account for retrogressions, as holds for either the Tychonic or the inverted Ptolemaic system, contrasts with the multiple motions within the standard Ptolemaic system, which just happen to coincide—the always parallel positions of the vectors discussed in connection with figures 7.2 and 7.3. For Ptolemy, the Sun plus Mercury and Venus move together, but the epicycles of the remaining three planets all turn at rates unique to each but such that the position of the planet on its epicycle always matches the position of the Sun relative to the Earth. In either the inverted system or in the Tychonic, however, a single motion carries all the planets along with the Sun. For these systems (and much more simply for the Copernican)[8] a common motion carries the annual component of motion of the Sun and all the planets.

Further, in the standard Ptolemaic system there is no explanation of why two planets stay close to the Sun while the other three orbit independently of the Sun. That is just the way the world is. But in the inverted system, exactly as in the Tychonic, there is a simple explanation. If the secondary orbit (the p-orbit for all five planets after the inversion) is larger than the Earth-Sun distance, then the planet as seen from the Earth will appear to move independently of the Sun. Conversely, if the p-orbit is smaller than the Earth-Sun distance, then it appears to move with the Sun. This contrast is now only a superficial appearance: all the planets in this inverted system, as also holds in the Tychonic system, move in the same way.

On the other hand, an advantage of the standard system seems to be that all the epicycles turn in the same direction as the Sun and the deferents turn. This holds in the inverted system only if the Ferris-wheel linkage motion keeping the orientation of the epicycle fixed in space is made explicit [7.3]. Otherwise, since in the inverted models the speed of the inner circle is faster than that of the outer, the planet has to be postulated to move backward on its epicycle.

Look at figure 7.8, which shows Mars three months after the setup depicted in figure 7.5a. Note that the radius of the epicycle in the standard configuration (the short horizontal line labelled r_{st}) moves clockwise as the figure is drawn, as does the radius of the deferent (R_{st}). After inversion, the radius of the epicycle (now the longer line, R_{inv}) seems to move opposite to the rotation of r_{inv}. That happens because the linkage [7.4] has been left implicit.

Consequently, an advocate of the inverted system faces a choice between complicating his system by making the linkage condition explicit, or of confronting the oddity of why the outer planets move backward on their epicycles. But once this difficulty is faced, an additional and striking new advantage to the inverted system appears. For when the linkage is made explicit, the periods of the p-orbits fall into an order they lack in the standard Ptolemaic system. In the standard Ptolemaic scheme Venus has a longer period than the Sun though on the nested-spheres logic it must come inside the Sun. But once the Ferris-wheel linkage is made explicit, the period of Venus drops from 584 days (as it seems until the motion of the s-orbit has been untangled from the intrinsic period of the p-orbit) to 225 days.

So we get the kind of sequence of discovery that ordinarily powerfully supports a novel idea. At first cut the inverted system leads to a bizarre consequence: the orbits of Mars, Jupiter, and Saturn go backward. It is necessary to postulate the explicit Ferris-wheel linkage to make better sense of that. But once that linkage is specified to make sense of what is happening with the outer planets, it turns out to resolve also what had been an anomaly for the inner planets (that the period of Venus is longer than that of the Sun).

Since the Tychonic and inverted Ptolemaic systems differ solely in how they are scaled, the same linkage is also required for the Tychonic system if the oddity of orbits that go backward for the outer planets is to be avoided. But the offsetting advantage is not so striking, since there is no Tychonic reason that the period of the Sun should fit neatly between the periods of Venus and Mars. On the contrary, the neat sequence of periods that appears, Kepler is able to claim, "shouts out loud" that the world is Copernican.[9]

There is a further advantage of the inverted system. The lopsidedness of the Tychonic configuration proves troublesome to everyone who contemplates it. Tycho persuaded himself (mistakenly, as I have just argued) that the problem is an incompatibility with solid spheres, so that once he abandoned the traditional commitment to solid spheres he felt that the system was justified. But even today everyone who looks at the Tychonic diagram feels uncomfortable with it, long after solid spheres are an almost forgotten curiosity. It does not accord with our physical experience of how things move in the world. A Tychonian would deny that earthly physical intuitions are relevant to what goes on in the heavens, but on the record (and consistent with the habits-of-mind view) that logically reasonable rebuttal is not enough to make the system intuitively comfortable. The inverted system has no such problem.

Still another advantage of the inverted system as compared with the Tychonic is that, under the Tychonic system, the traditional explanation of the complicated motions of Mercury (that it is closest to the Earth, and hence subject to the vagaries of terrestrial motions which cause even greater complication for the Moon) is lost.[10] For in the Tychonic setup as in the Copernican, Venus comes closer to the Earth than Mercury does. In the inverted system, as in the standard Ptolemaic setup, Mercury comes next to the Earth, then Venus, then the Sun, as their periods (using the explicit Ferris-wheel linkage) of 88, 244, and 365 days imply they should.

Overall, then, inversion gives us a system both less radical and more plausible than the Tychonic system over which Tycho had to fight bitterly to defend his claim to priority. But no contemporary of Tycho proposed it. And this system also exhibits superior internal coherence compared to the standard Ptolemaic system: the relation between the order of the planets and their periods is more convincing, a simple explanation of the apparent contrast in motion between inner and outer planets becomes obvious, and we no longer need appeal to some unmotivated coincidence that makes the s-orbits of the outer planets turn at whatever rate keeps the position of the planet on its epicycle parallel to the Earth-Sun line. But no pre-Copernican astronomer ever proposed the inverted system either. Why was this system never considered by anyone?

8.4

Obviously, no one could come to *believe* in the inverted alternative until someone had come to notice it. No one, however, in fact seems to have noticed it. So the puzzle is not why the merits of the inverted system won it no support, but why no one ever seems to have noticed the inverted possibility at all. This cannot be because of any logical difficulty of the path to discovery of this system. Apollonius had shown the possibility of inverting epicyclic models centuries before Ptolemy, and Ptolemy specifically mentions the possibility (*Almagest* XII.2). Oddly, Ptolemy (or perhaps some later editor, for this seems to be merely a blunder) limits the possibility of inversion to the outer planets. But since the inverted system in fact requires inversion only of the outer planets, that would not discourage the move under discussion here. So Ptolemy, and also every careful student of Ptolemy over the next 1,400 years—which is to say every astronomer of any significance—was logically aware of the possibility of inverting the outer planets. Why did no one see it? Why, even after Copernicus, should all astronomers unable to follow Copernicus have *uniformly* favored Tycho, with (so far as we know) not even a bit of consideration of the inverted Ptolemaic alternative?

The cognitive explanation I want to suggest turns on the aspect of the habits of mind argument alluded to in the opening paragraph of this chapter. Habits of mind that make it hard to see things in one way will also make it hard to avoid seeing some alternative view.[11]

A person who takes the trouble to plot the movements of a planet against the background of fixed stars generates paths like those of figure 7.1. Although the immediate perception of the heavens is of a dome with lights on its surface (like the ceiling of a planetarium), an astronomer prompted by clues such as those mentioned in section 7.2 can come to see the planets as riding on epicycles, with the varying loops generated by seeing the epicycles at varying distances and varying angles. Psychologically, we have no reason to suppose that what happened for a Ptolemaic astronomer differs in any way at all from what happens as a modern scientist becomes familiar with oscilloscope tracings of what he comes to *see* as electron paths.

In section 7.5 we saw that inverted models yield exactly the same traces as their standard Ptolemaic counterparts. But our experience in the world is almost always of little things riding on big things (as in the standard models of figures 7.2 and 7.3), not the converse (as in the inverted models of figures 7.9 and 8.1). And that is enough to explain why, when astronomers considered paths like figure 7.1, they very naturally saw the loops in terms of small-on-big, not the reverse, just as when chemists prior to Lavoisier watched a fire burn, they saw the process in terms of something escaping, not something being added to the fuel. It takes the exercise in geometry reviewed in section 7.5 to show that inverted models will produce exactly the same pattern of observations as the more comfortable small-on-big models.

Since for Mercury and Venus the p-orbit must be smaller than the s-orbit (to match observations), and the converse for Mars, Jupiter, and Saturn, the usual arrangement (p-orbit on s-orbit for Mercury and Venus, s-orbit on p-orbit for Mars, Jupiter, and Saturn) is going to look right and the alternative look wrong, even though mathematically there is no reason whatever to prefer one over the other. It is only when all the individual models are put together to form a *system* that big-on-small models (for the outer planets) have a chance to be seen as looking more reasonable.

But Ptolemy himself was working in a tradition already several centuries old by the time he did his great work. So although there is a strong sense in which the inverted scheme is more coherent, by the time a strikingly effective cosmological system finally was a realistic possibility, the astronomer's now well-entrenched and entangled habitual way of seeing things would have been a barrier to thinking things through in terms of an inverted system. The possibility of any such insight could only come after Ptolemy, apparently late in his own career (there is no hint of it in the *Almagest*), had constructed the nested-spheres system of the world, which then dominated astronomy for the next fourteen centuries. By the time Ptolemy might have seen the virtue of the inverted alternative, he—and through him, later astronomers—were very solidly entrenched in the more immediately comfortable small-on-big setup for individual planets.

8.5

But since the "coincidence" argument appears so strong in the sixteenth century, why would it not have provided an anomaly strong enough to push earlier astronomers toward consideration of the inverted system? The answer turns on a point that arose in the Lavoisier discussion (chapter 4, note 6) in connection with the claims that phlogiston helped to explain the shiny, malleable, and other common properties of metals. The plausibility of an argument is always in part context-dependent. How plausible an idea is to a person will vary with what else is believed or what else might be hard to avoid if this belief was rejected.[12]

If there were no offsetting disadvantages to the inverted setup, it would be astonishing if its greater coherence relative to what became the standard Ptolemaic system failed to prevail. But as I have already stressed, the *systematic* advantages of the inverted scheme could not be recognized until Ptolemy had developed things to the point where any *system* at all could be envisioned, which was not until several centuries after Apollonius worked out the basic theorems. Small-on-big was not just cognitively easier (because closer to endemic intuitions about how things work) but had become entrenched and entangled with the practice of mathematical astronomy in the way I have discussed at length in other contexts. Once the standard model had taken hold, the taming of anomalies [3.4] would operate to make the coincidence and other relative disadvantages of the standard scheme effectively invisible. That each planet had an orbit that exactly tracked the motion of the Sun would come to be seem as a inherent feature of the world. Like a disaster or an assassination, which eventually comes to seem inevitable, a world without orbits tied to the Sun would seem somehow to be missing something it should have. By the time of Ptolemy, and perhaps for a long time before Ptolemy, everyone expert in astronomy was well trained to see things in the standard way, which would inevitably include intuitive training to see the coincidence of solar-tuned motion not as a coincidence at all, but as an intrinsic property of the planets. To be such a body was to have an s-orbit. Each planet has an s-orbit whose *function* was to move always in whatever way would track the Sun. That each planet had such an s-orbit would no more be seen as a coincidence that could raise doubts about the system than that every man you see walking in the street is wearing pants is a coincidence that puzzles you when you walk down that street. It is just the way things are.[13]

8.6

But consider now the situation around 1600, when Kepler's language [8.2] so clearly implies that the coincidence argument was a very strong one against Ptolemy. How did the coincidence argument become strong after being tamed for so many centuries? Given that it had some-how become strong, so that the standard Ptolemaic beliefs were falter-ing, why did the inverted possibility still remain out of sight? Why did conservative astronomers instead move uniformly to what we now call the Tychonic system, without (so far as the record shows) ever even noticing the inverted possibility?

The answer to the first question now seems easy, since it can be seen as just another illustration of the point about the context-dependence of plausibility. Copernicus proposed a system that did more than consolidate the distinct solar motions into a single motion affecting six bodies (the Sun plus all the five planets at once, as is done also in the Tychonic or inverted Ptolemaic systems). In Copernicus's system loopings of the planets are not merely consolidated in the way that the inverted or Tychonic system consolidates them compared to the stan-dard Ptolemaic system: they disappear entirely. Further, in the Coperni-can scheme the linkage motion disappears (or, equivalently, the need to assign periods for the outer planets that move backward [8.3]).[14]

On the one hand, the counterintuitive notion that has to be swallowed with the Copernican argument is very difficult: the solid Earth has to be seen as an object flying through space. On the other hand, the simplification involved is far more striking than that involved in going from the standard Ptolemaic to the Tychonic or to the inverted Ptole-maic system. And it is accomplished without the impoverishment of mental imagery that occurs with the inverted scheme in either its Ptole-maic or Tychonic version. It is not hard at all to come to see the re-trogressions as due to the parallax effects of the Earth's own motion, but hard to see (with mental imagery) the Tychonic or inverted Ptolemaic retrogressions. When striking new evidence or (as here) striking new arguments reopen a choice that had been wholly settled, a *tamed* anomaly (here the coincidence of s-orbits) can become salient again.

This still leaves the puzzle of why the conservative response was to move to the inverted Copernican system proposed by Tycho and various rival claimants to invention of this system, not to the in-verted Ptolemaic system I have been describing. In particular, why even at this stage did the inverted Ptolemaic possibility continue to go wholly unnoticed?

8.7

Suppose you had come to see things in the Copernican way, and were held back only by the difficulty of believing the Earth is actually a planet. In other words, your habits of mind are no longer the nested-spheres habits that on the barrier argument are *constitutive* of the Ptolemaic paradigm. If we had access to a live Tychonic astronomer, there would be a concrete test available for that. Let the person imagine his view of the heavens and watch what happens as a retrogression occurs for some planet.[15] The reader should try it, taking care to attempt to see the physical planet moving in space, not just diagrams. Since we are all Copernicans, even if you are familiar enough with the Ptolemaic diagrams to be able to comfortably see the retrogressions in terms of how the diagrams work, you will still have trouble imagining the actual planet behaving in that, as Kepler called it, pretzel-like way. In particular, it would be difficult to do that once you lost what must have been the very secure and concrete sense a Ptolemaic astronomer would have had of the epicycles as physical objects in space. In terms of the language I used in *Patterns* (chapter 12), it would be difficult to do once the solidly entrenched nested-spheres sense of the world had been displaced by the golden-chain sense of things that structure the cosmology of the Copernican world.

As the moves apparatus of chapter 7 shows, going from the standard Ptolemaic to the inverted Ptolemaic scheme is a matter of one step: invert the outer planets. The sequence and dimensions of the planetary spheres remain just as they were in the standard system. Similarly, going from the Copernican setup to the Tychonic is also one step: slide the Sun back to its geocentric orbit, which (reversing the slide move) will carry the orbits of the planets along with the Sun.

So suppose that men like Tycho in fact already had lost the nested-spheres sense of things, since that had been displaced by the Copernican golden-chain sense. Then the easiest way to envisage the inverted Ptolemaic system would be to go backward through the moves discussed in section 7.5, moving to the inverted Ptolemaic system by way of the Tychonic system. So you are bound to at least consider that Tychonic alternative. And you would not feel uncomfortable at the loss of the nested-spheres symmetries of the Ptolemaic system. You have already lost that. The Tychonic system is as incompatible with nested-spheres as the Copernican is. Rather than feel uncomfortable with its loss, you will feel uncomfortable with the loss of the golden-chain relations if you now return to a Ptolemaic nested-spheres scheme. The inverted Copernican system is closer to your immediate intuitions than the inverted Ptolemaic system is. If you did want to get to that inverted Ptolemaic system, you would find yourself doing it not in one step away from Ptolemy but in two steps back from Copernicus—the first being the step to the Tychonic system. And there you would find a further powerful inhibition against going back that second step.

For an important point to notice about the Tychonic system is that the intrinsic difficulty of coming to believe in the reality of epicycles (in particular the difficulty of sustaining that belief once Copernicus had offered an alternative) does not *logically* affect the inverted system relative to the Tychonic, since the p-orbits are epicycles on either system.[16] Cognitively, though, things are different. In the Tychonic system the deferents of the planets coincide with the orbit of the Sun, so although they are there on the logic of the system, you can't see them when you look at a diagram. At least if you only think of things in terms of diagrams (and don't try to envisage what the planets must be actually doing in space), then the Tychonic diagram looks almost as clean of epicycles and deferents as the Copernican. Since Tychonic astronomy throughout its few decades of life was always parasitic on Copernican analysis (it developed no independent technical apparatus of its own), a Tychonic astronomer could very comfortably learn to evade thinking directly about how the system might work.

We would have a mystery indeed if conservative astronomers went from seeing the world in the Ptolemaic way to seeing it in the Tychonic way, entirely overlooking the inverted possibility. But we have no mystery at all if, forty years after Copernicus's book appeared (for that is when this tendency appears), we find that astronomers—who by this time would be learning of the Copernican ideas at the start of their careers, before solidly entrenched Ptolemaic habits of mind could take hold—no longer have intuitions dominated by the Ptolemaic nested-spheres habits of mind. In that state, prevailing habits of mind would facilitate (not inhibit) seeing the inverted Copernican system of Tycho as a geocentric alternative to Copernicus, and they would become a barrier to seeing the alternative conservative possibility of the inverted Ptolemaic system.

The blindness to the inverted Ptolemaic possibility the record reveals makes no sense either logically or psychologically unless such a shift had in fact occurred. But we have entirely different grounds (*Patterns*, chapter 13) for in fact supposing that that shift had occurred. If by the 1590s nested-spheres habits of mind had not been lost—not because astronomers who once had that sense of things had lost it, but because a new generation of astronomers were on hand who had never become entrenched in nested-spheres habits of mind—it would be bewildering that belief in Ptolemy would fade without a fight after dominating its field for 1,400 years, as clearly it had even before Galileo's telescope finally produced empirical evidence against it. On the other hand, if nested-spheres habits had faded, then that also explains why the "new" Ptolemaic system we have been considering, which logically should have had great appeal to conservative astronomers, never was so much as noticed. If astronomers entrenched in nested-spheres habits of mind could miss the heliocentric possibility for fourteen centuries—as on the record they did, presumably for the cognitive reasons reviewed earlier—it is hardly a puzzle that their successors could miss the inverted Ptolemaic possibility for the few decades when any astronomer of consequence defended a Tychonic world.

Nine

A Copernican Detective Story

The Copernican case is not only an obvious candidate for the barrier analysis, but also a particularly good case to illustrate the uniqueness claim. It is very plainly an example of a situation in which some very stubborn cognitive difficulty must have blocked a discovery. Copernicus used no essential information not provided by Ptolemy. The crucial steps needed to transform the Ptolemaic scheme into a heliocentric equivalent are not difficult, as we have seen in chapter 7. But this remarkable argument—one so intrinsically neat that it would never have been forgotten even if the telescope had shown it to be untenable instead of irresistible—lay on the table for fourteen centuries with no one able to see it.

There are two candidates for barrier: one obvious (the difficulty of seeing the Earth as a planet flying through space) and another less obvious (the nested-spheres sense of the structure of the world emphasized in chapters 7 and 8). So we have a concrete example of the situation described as an abstract possibility in chapter 3, where there are two (or more) candidates for barrier. On that earlier argument, we expect that nevertheless it would almost necessarily turn out that just one potential barrier plays a unique role in the emergence of a revolutionary idea.

Three possibilities arose in the earlier discussion, each of which effectively left a unique barrier for the episode. (1) There might be striking breakthrough effects from breaching one barrier, but the second barrier remains stubbornly entrenched. In that case, the breakthrough would be remembered as a remarkable discovery, and the breach of the second barrier, when it came, would also be so remembered: as with publication of the Copernican discovery, then Kepler's abandonment of uniform circular motion and, with that, abandonment of the entire Ptolemaic apparatus, seventy years later. (2) If there are no breakthrough effects, then the episode would not be well-remembered at all, since no striking discovery was made—or even more likely, the anomaly that made getting past the barrier feasible would come to be tamed, with the barrier remaining in place. Only with hindsight might someone notice the unrealized significance of breaking a formidable barrier, and point to the occasion as a precursor of a famous discovery. (3) Finally, the breach of one barrier might yield sufficiently striking breakthrough effects to make the second barrier—perhaps originally more formidable—now vulnerable, as I have argued for the case at hand.

9.2

In chapter 7 I tried to show why the major epicycles could be *seen* in the mind of a Ptolemaic astronomer with the conviction that a conductor can hear the music of the score she is reading.[1] For a well-trained Ptolemaic astronomer, there would be nothing artificial or even conjectural about the way the epicycles served as the determinants of the spatial structure of the Ptolemaic universe. On the contrary, the neatness with which the epicycles served this role as well as accounting for anomalies of motion gave a coherence and hence a powerful sense of conviction to the overall system [7.4]. The distances and sizes of the planets were computed in a way that made no sense unless the epicycles were real, space-filling objects, governing the sizes of snugly nested spheres that filled the universe.

The nested spheres became even more intricately entangled in the way the world was seen by astronomers through the analogous arrangement of the spheres of the Aristotelian elements, and then the melding of both with Christian theology and with the geography of the Earth, as discussed in detail in *Patterns* [12.7]. The nested spheres came to extend from Jerusalem at the very center of everything out to the farthest reaches of the universe. Well-entrenched habits of seeing the world in the nested-spheres way would yield fluent use of these ideas and a confident sense of how everything fits together in a way too coherent to be doubted.

To a modern reader this Ptolemaic way of seeing the world

easily seems merely a horribly compounded blunder. In terms of our own habits of mind, the nested-spheres idea makes no serious sense, and it is only with some effort and practice that we can begin to see how compelling it might be. But until the Copernican argument had been spelled out, and (on the record) even for several decades thereafter, the nested-spheres idea was more easily seen as obviously right than as possibly wrong. It provided a powerful tacit sense of how the world worked, not simply an explicit belief, which might be vulnerable to counterarguments. Given this deeply entrenched sense of things, the explicit nested-spheres principle became self-evident, so that until some alternative way of seeing things with comparable appeal had become familiar, even a good counterargument would be seen as only a clever trick, not as a reason to seriously doubt what everyone knew.

As developed in detail in *Patterns* (chapter 12), the critical event was the publication in 1507 of an enormous map celebrating the astonishing idea that, in gross violation of nested-spheres intuitions, a new continent had been discovered on the back of the Earth.[2] We have the curious coincidence that after 1,400 years, the heliocentric possibility was seen within, at most, seven years of publication of Waldseemüller's map. Columbus had died a year earlier, still believing he had reached the eastern limits of Asia.

9.3

The heliocentric discovery was impossible so long as either of the two well-entrenched habitual senses I've mentioned governed intuitions (central-Earth and nested-spheres). For the Earth-as-center way of seeing, this holds immediately: if the Sun is the center of motion and the Earth is in orbit, then it can't be that a fixed Earth is at the center. But the rigid incompatibility of the heliocentric idea holds as well for the nested-spheres sense of the world. The major epicycles disappear in the heliocentric scheme. The nested-spheres world, where those epicycles determine the spacing between the heavenly bodies, collapses.

Hence so long as intuitions were governed by *either* nested-sphere or central-Earth senses, the normal response to a heliocentric intuition could only be that it looks crazy, makes no sense, is obviously wrong even if you can't say why it's wrong, as with the response to a clever proof that $2 + 2 = 5$.

But although the central-Earth intuition could not be challenged without also challenging the nested-spheres sense, the converse does not hold. The Tychonic setup collapses the nested-spheres world just as the Copernican scheme does, though without challenging the central-

Earth intuition. Further, reaching a Tychonic setup requires only the invert and scaling moves [7.6], neither of which even needed to be invented by a discoverer, since both are described in the *Almagest*. So while there is no way to attack the central-Earth aspect of Ptolemaic astronomy without simultaneously attacking the nested-spheres, the converse does not hold. Further, the Tychonic setup does produce breakthrough effects, as will be noted shortly. Hence, cognitively, it looks both easier and more promising to start with a challenge to the nested spheres than with an immediate challenge to the central Earth. It looks easiest of all to start by moving the epicycles of Venus and Mercury to center on the Sun, which violates the nested-spheres intuition but does not require challenging yet another very well entrenched habit of mind—the small-on-big sense of how things move which played the central role in chapter 8. For it is only the outer planets that need to be inverted to reach the Tychonic setup.

Novel and interesting relations appear with the shift of the inner planets to heliocentric orbits: in particular, a wholly unambiguous ordering of Mercury and Venus appears in place of the Ptolemaic plausibility argument, which in fact was most vulnerable here, since (in contrast to the outer planets) the ordering was not uniquely determined by increasing period.[3] But the heliocentric ordering of Mercury and Venus is the opposite of Ptolemy's in terms of closeness to the Earth. There was no strong reason for an astronomer to believe this is better than Ptolemy's setup, or even to take it very seriously as a physical possibility. But an astronomer who could break free enough of nested-spheres habits to find that this first, partial heliocentric move leads to something interesting would easily be led on further—in particular to consider the possibility, also lying in the wings for 1,400 years, of inverting the outer planets so that they too could be put into heliocentric orbits. After all, once we are primed to see the heliocentric possibility, it is not logically difficult to think that those planets come closer to the Earth when the Earth is between the Sun and the planets, because they are in orbit around the Sun.[4] A salient but unexplained feature of the Ptolemaic world here becomes a necessary feature of the alternative model.

Overall, it is not hard to sketch an account in which a competent astronomer endowed with a strong sense of curiosity would be led further and further into interesting possibilities that are revealed as the planets (but not yet the Earth) are treated "as if" heliocentric, soon reaching a point where it becomes tempting to make the decisive move of seeing what things look like if the Earth itself is treated as a planet. The new element in the situation (in the immediate aftermath of the startling news embodied in the Waldseemüller map) is that the very first step—moving Venus to a heliocentric position—might be taken without it looking immediately absurd and not worth thinking about. For at just this moment the countervailing thought that the nested-spheres sense might be somehow wrong, or at least not absolutely right, was peculiarly available.

Once a person had gotten beyond the nested-spheres barrier, he might be able to go further and see that, if the Earth itself were put into heliocentric orbit in the large space left between Venus and Mars, then the major Ptolemaic epicycles become images of the Earth's own orbit around the Sun. Hence it is *of course* the case that a major element of each planet's motion is linked ("as if by a golden chain," as Copernicus's disciple Rheticus put it) exactly to the Sun's apparent motion around the Earth. Once that simple relation came clearly into view, other old and long-tamed anomalies (that the huge Sun orbits the tiny Earth, and others) could be seen in a fresh way.

So we have here a very striking case where getting past one barrier could yield breakthrough effects, which then ease the difficulty of getting past what had previously been a second, cognitively even more difficult barrier.

9.4

In *Patterns* I provide a good deal of detailed evidence to support this nested-spheres analysis of the Copernican discovery. I also sketched an account of the most plausible route to the discovery, which would proceed, as just suggested, by first moving the epicycle of Venus and then the epicycles of other planets to a heliocentric position while leaving the central position of the Earth initially unchallenged. And we have some serious, though not compelling, evidence that Copernicus indeed started with the particularly salient (for this analysis) move of Venus to a heliocentric orbit. But putting the moves of chapter 7 to work once again, a curious but striking further bit of evidence can be seen in a few notes bound with Copernicus's personal copy of the Alphonsine tables (Swerdlow 1973, 475). It provides what seems to be a quite unambiguous piece of direct evidence that Copernicus indeed followed what on the barrier argument would be the easiest route to the heliocentric discovery; that is, he started with Venus, which would involve breaking with the nested-spheres habit of mind first, but in a way that would yield striking breakthrough effects without yet challenging the sense of the Earth as the solid center of all motion.

We happen to know enough of Copernicus's work just prior to the period of the discovery to expect that a salient focus for him would have been an attempt to resolve a long-standing anomaly involving Venus. This turned on the absence of any significant variation in the brightness of Venus as a function of its position relative to the Earth. Since in Ptolemaic models (and as the connections developed in chapter 7 assure, in the Copernican model as well) Venus varies by about a factor of six in distance, that was a puzzle. This Venus puzzle was cited by Osiander in his notorious preface to *De Revolutionibus,* warning its readers that no astronomical system should be taken as physically true.[5] Copernicus was able to resolve a similar puzzle for the Moon, by defining a mechanism in which the Moon's motion was captured without the impossibly large variation in distance from the Earth implied by Ptolemy's model.

So it is a very reasonable conjecture, although it is only a conjecture, that Copernicus would follow his work on the Moon by considering alternative models of Venus, adapting the technique he had successfully used to achieve a striking improvement over Ptolemy's model for the Moon. He specifically says (in opening the *Commentoriolus* [Swerdlow 1973]) that he was led to consider "alternative arrangements of circles." If that exploration included moving the epicycle of Venus to center on the Sun (instead of lying between the Earth and the Sun), then that would provide just the step needed: it involves a breach of nested-spheres habits of mind that initially involves no challenge at all to the sense of the Earth as at the center. And it is a breach motivated by a problem Copernicus was likely to find salient. And it turns out that indeed there is evidence that Copernicus proceeded in just this way.

9.5

In *De Revolutionibus* (1543, I.10) Copernicus suggests that he may have started by moving Venus and Mercury before proceeding to put the outer planets into heliocentric orbits, and that only then did he consider moving the orbit of the Earth. But the suggestion is indirect, in the sense that it is framed in terms of how a reader might most easily see the logic of the heliocentric move. Kepler (Jardine 1984, 65) interpreted Copernicus's suggestion about how a reader might see the point as a description of how Copernicus himself did see it. But since Kepler's remark comes half a century after Copernicus's death, it seems only a conjecture, though it is the conjecture of someone living at a time when it was still a radical thing to come to the belief that the Earth is a planet. So it is the conjecture of someone who would have more basis than we do to judge how such a seemingly absurd idea (as Copernicus himself characterized it at the outset of *De Revolutionibus*) could come to be believed.[6]

It turns out, however, that a much more direct bit of evidence

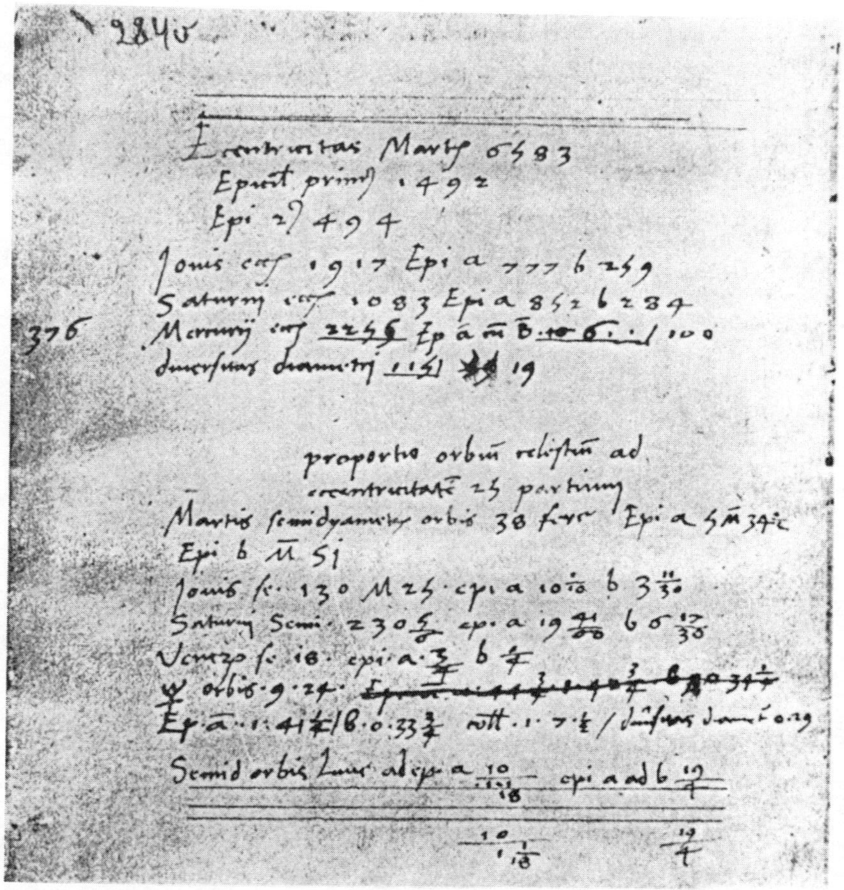

Eccentricitas Martis 6583
Epiciclus primus 1492
Epi[ciclus] secundus 494
Jovis ecce[ntricitas] 1917 Epi[ciclus] a 777 b 259
Saturnij ecce[ntricitas] 1083 Epi[ciclus] a 852 b 284

376 Mercurij ecce[ntricitas] 2256 [2259?] Epi[cicyckus] a cum b·10·6· ·/100
diversitas diametrj 1151 $\overline{59}$ 19

proportio orbium celestium ad
eccentricitatem 25 partium

Martis semidyameter orbis 38 fere Epi[cyclus] a 5 $\overline{M}$ 34
Epi[ciclus] b $\overline{M}$ 51
Jovis se[midiameter] 130 M 25·epi[cyclus] a 10$\frac{1}{10}$ b 3$\frac{11}{36}$
Saturnij Semi[diameter] 230$\frac{5}{6}$ epi[cyclus] a 19$\frac{41}{60}$ b 6$\frac{17}{36}$
Veneris se[midiameter] 18·epi[cyclus]·a·$\frac{3}{4}$ b $\frac{1}{4}$
☿ orbis·9·24·Epi[cyclus] a 1·44$\frac{1}{4}$·1·42$\frac{3}{4}$·b 0 34$\frac{1}{4}$
Epi[ciclus] a·1·41$\frac{1}{4}$/b·0·33$\frac{3}{4}$ coll[ligunt?]·1·7·$\frac{1}{2}$·/ diversitas diametri 0·29

Semid[iameter] orbis Lune ad ep[cyclum] a $\frac{10}{1\frac{1}{18}}$ epi[cyclus] a ad b $\frac{19}{4}$

$$\frac{10}{1\frac{1}{18}} \qquad \frac{19}{4}$$

Fig. 9.1. Copernicus's notes calculating the heliocentric distances. The upper set lists a variable extracted for each of the planets except Venus from
his copy of the Alphonsine tables, with which these notes were bound. The
lower set gives the heliocentric distances for all the planets (including Venus). As explained in the text, going from the Ptolemaic values of the Alphonsine tables to the heliocentric values is actually very simple. But there
is the puzzle of why Copernicus did not also need a value for Venus in the
upper set of numbers. The simplest explanation is that Copernicus had
started by working out the heliocentric possibility for Venus, and only then
was led to work out the distances for the full set of planets. Reprinted with
permission from N. Swerdlow, "The Commentariolus of Copernicus," *Proceedings of the American Philosophical Society* 117 (1973): 428.

has also survived that indeed Copernicus started with Venus, as the barrier argument encourages us to suppose. The notes Swerdlow discovered bound with Copernicus's personal copy of the Alphonsine tables are shown in figure 9.1, along with Swerdlow's transcription of Copernicus's Latin script. The upper set of numbers labelled *ecce* show the maximum value (at mean distance) for the sine of the angle that an astronomer could observe between the computed center of the Ptolemaic epicycle and the actual location of a planet. The reason for constructing a table of these values is to let an astronomer (or, then usually the same thing, an astrologer) compute the position of a planet relative to a future position of the center of its epicycle. Copernicus is just running his finger down these tables and picking out the maximum value.

But as explained in the note appended here, a person with that number in hand could immediately compute the heliocentric distance of the planet, even though the numbers in the tables were computed by geocentric models to make geocentric predictions.[7] All that is required is that the s-orbit for each planet (scaled in the Ptolemaic system by the nested-spheres principle) be rescaled to equal the Earth-Sun distance. As the label on the lower set of figures shows (noting that all distances are scaled to a common Earth-Sun distance set at 25,000 units), that is just what Copernicus was doing.

Because of the relation shown by the moves of chapter 7, a single multiplication or division takes Copernicus from a number he can read from his Ptolemaic tables to the heliocentric distance for a planet. There is no reason to wish we had the intervening details that let Copernicus go from the geocentric numbers in the upper table to the heliocentric numbers in the lower. Aside from a bit of scratch paper, there are no intervening details: it is a one-step operation. That these brief notes in the margin are really all there is to working out the heliocentric distances is a vivid indication of how short the basic step is *logically* from geocentric to heliocentric astronomy, and consequently how striking is the delay of fourteen centuries before anyone could see the heliocentric possibility.

9.6

Now notice that the parameter taken from the tables for Mercury, Mars, Jupiter, and Saturn is listed as "ecce" (for "eccentric"), which Swerdlow points out makes sense only if Copernicus was thinking of inverted models when he made this list.[8] We would have (for the outer planets, for which the Tychonic models require this inversion) the big-on-small setup discussed at length in chapter 8. This fits the route to seeing the heliocentric idea that Copernicus mentions in *De Revolutioni-bus* I.10, where the steps consist of moving the epicycles of the inner planets to center on the Sun, then moving the outer planets also into heliocentric orbits. The technical step required to make that move is to invert the models into big-on-small form, as described in section 7.5.

There is no indication that Copernicus ever saw the Tychonic positions as physically meaningful, or that anyone else did either until the 1580s (Gingerich and Westman 1989). But that arrangement had interesting mathematical properties, namely, those already discussed in connection with the inverted Ptolemaic system in chapter 8, plus a large empty space between Venus and Mercury into which a planet with a heliocentric period greater than that of Venus (0.6 years) and less than that of Mars (2 years) might fit, even if sufficient space to carry an orbiting Moon was also required. So it is not a surprise that Copernicus would eventually notice that there was such a planet.

9.7

But notice a peculiarity of Copernicus's notes. In the lower set of numbers, Copernicus gives the distances for *all* the planets. In the upper set of numbers, where he gathers the parameters for doing the calculations, only four of the five planets are listed. How could he have gotten the distances for five planets in the lower set of numbers when he gathered the parameters for only four? Why did he not need the parameter for the remaining planet—which happens to be Venus? I don't think any plausible explanation will be suggested other than that he didn't gather the parameter for Venus here *because he had already done the Venus calculation before he came to treat the other planets.* We seem to have caught a glimpse of Copernicus just at the moment when, having taken the Venus step, he was sufficiently struck by what he saw that he was moved to do the same for the rest of the planets.[9]

Once again the exceedingly scanty record available from the early sixteenth century happens to include scraps of information that make easy sense in terms of the barrier view of the discovery, but are puzzling coincidences otherwise. If wrong, the barrier account of the episode is accompanied by what adds up to a remarkable series of coincidences suggesting it is right.

We are also provided here with a striking example of how expertise in handling the technical content of a paradigm does not necessarily imply seeing the world, or even ability to see the world, in the way that ordinarily could not be escaped by the historical actors whose expertise is at issue. A person can be fluent in the theory but quite incapable of seeing things in the way that historical figures working within the Ptolemaic paradigm saw things. Similarly, historical actors with an expert command of the technical material can be blind to possibilities that seem obvious to us—here blind for 1,400 years to the logically simple Copernican, Tychonic, and inverted transformations of the Ptolemaic setup.

In the language I have been using, expertise always involves deeply entrenched habits, since fluent performance of any kind (whether physical or mental) always depends on that. But the habits of mind that are essential to a paradigm are not just those that any expert practitioner would automatically come to acquire. There is always an important element of what in the artificial-intelligence community is called "chunking," and which in the language I have been using would be treated as an aspect of the notion of "entanglement." The effect is to bundle together various details in some higher level all-at-once way of seeing a range of things. This is bound to be contingent in a path-dependent way on how the ideas developed and on interactions with other beliefs and habits of mind.

A modern expert in Ptolemy always comes to the subject with solidly entrenched habits of mind inconsistent with the central-Earth, and with no commitment whatever to the nested-spheres sense of things. So although a modern expert in Ptolemaic astronomy would eventually acquire some facility with nested-spheres thinking, there is no plausible way in which that modern expert could acquire the nested-spheres habits of mind in the deep and robust way that an actual Ptolemaic astronomer could hardly escape. Rather, the natural way for a modern scholar to gain a fluent command of Ptolemaic astronomy is by seeing the old theory as a transformation of current beliefs. We inadvertently illustrated just that point when we found the easiest way for us to grasp how Ptolemaic astronomy works was by explicitly using the linkage motion, which actual Ptolemaic astronomers consistently left implicit. The argument [8.7] about why the Tychonic alternative so wholly dominated the inverted Ptolemaic as a possible conservative response to Copernicus suggests that, even by the closing years of the sixteenth century, astronomers were no longer bound by nested-spheres intuitions, and apparently no longer even had ready access to them.

Hence in the sharpest possible contrast to having the habits of mind that were essential constituents of the Ptolemaic paradigm— habits that would make it extremely difficult to see the heliocentric alternative as making sense—a modern expert (and on the argument of chapter 8 perhaps even astronomers at the close of the sixteenth century) would acquire fluency in Ptolemaic astronomy, to the extent they do so at all, by seeing the details as transformations of heliocentric models. With effort, given that we know where we have to get to, we can work our way into seeing things in the superseded view we are trying to reconstruct. But what we must work at seeing, someone really committed to the older paradigm would see as immediately intuitive. Contrariwise, what a historical Ptolemaic astronomer could make no sense of, because it violated entrenched nested-spheres habits of mind, our modern expert can hardly see as anything but transparently reasonable.

Hence it becomes possible to understand how modern writers on Ptolemaic astronomy (Price 1959, Neugebauer 1968, and many tertiary accounts dependent on these) have sometimes so vehemently expressed their impatience with talk of a Copernican revolutionary discovery. For Price and Neugebauer, Copernican astronomy was best seen as a somewhat more complicated variant of Ptolemaic astronomy. Construed merely as a technique for modeling the angular motions of heavenly bodies ("saving the appearances"), that is a reasonable assessment of what Copernicus accomplished. But it is not a view that would have plausibly made sense to a sixteenth-century astronomer.[10]

Neugebauer, in particular, was well aware of the central role that the nested-spheres principle played in the Ptolemaic tradition. But he regarded that as just a "philosophical" diversion and not really what astronomy was about, hence his patronizing attitude toward the work that overturned a position not seriously challenged for 1,400 years and that convinced Galileo and Kepler that the solid Earth is a planet flying through space. In all this, it seems to me, we can see very strikingly what an enormous difference there can be between being an expert in Ptolemaic astronomy and being a Ptolemaic astronomer, actually seeing the world from within the Ptolemaic paradigm.

Indeed, if Ptolemaic astronomers did, or could, see things as a modern expert in Ptolemy sees them, free of the entrenched habits of mind that guided them, it would be an absolute mystery how fourteen centuries could slide by before some astronomer finally noticed the heliocentric possibility.

Ten

Ptolemaic Belief in a Copernican World

Introducing a volume of papers on scientific realism Leplin (1984) sums up the skeptical view this way:

> Whatever continuity may be discerned in the growth of empirical knowledge, theoretical science has been radically discontinuous. Scientific views about the ultimate structure and lawlike organization of the world have frequently been overthrown and replaced by incompatible views. Much of this discarded science was, for an appreciable time, eminently successful by the standards we employ in assessing current science. The inference seems inescapable that the evidence available to support current science is by nature unreliable and systematically underdetermines what ought to be believed about the world beyond our experiences.

A salient test case is provided by the Ptolemaic material we have been reviewing, since here a theory that was eminently successful, for a very appreciable time indeed, was radically overthrown when challenged by an amateur astronomer, working alone in his self-described "darkest corner of Europe."

Another perspective that makes this case salient: a defender of Kuhn's remarks about scientists with rival paradigms living in different worlds ordinarily needs to insist on a reasonable reading of the point. A new view ordinarily changes only a piece of the world the scientist sees, so that even a novelty on a very grand scale (like Darwin's) would not warrant a pedantic reading of the "different worlds" claim. But here views of the broadest possible scope, established for many centuries, were discarded and replaced by a radically altered ontology. The transition cannot be explained by supposing that the science at issue (astronomy before Copernicus) was in such primitive condition that we needn't expect successive theories to be tightly bound by common constraints; nor is there any story to be told here about the theory dependence of observation. We have a case in which successive theories of a well-developed science, dealing with the same observations, making the same predictions, nevertheless reach radically different pictures of "what is really there."

Figure 7.6 shows "what is really there" in the Ptolemaic scheme. The black "sandwich" (carrying the deferent) turns east to west every twenty-four hours; the deferent (carrying the epicycle) turns west to east with the period of the planet or following the annual motion of the Sun (for outer and inner planets, respectively). The epicycle turns (carrying the planet), with the appropriate motion to keep the position of the planet on its epicycle parallel to the Earth-Sun line, or with the intrinsic period of the planet (again for outer and inner planets, respectively), as described in detail in chapter 7.

A Ptolemaic astronomer would think very comfortably in these terms, and he would have many occasions to notice how beautifully those models capture the actual motions of the planets. He would become very fluent in seeing this setup in the mind's eye as he thought about how the world works. He would come to see the epicycles moving through space in particular ways that show themselves to an observer through the varying trajectories and spans like those illustrated in figure 7.1. His situation with respect to the epicycles would be like that of later astronomers, using telescopes, for whom the changing appearance of Saturn demonstrated the reality (however implausible a priori) of thin concentric rings around the planet, seen under varying perspective.[1]

In the Copernican world, these epicycles disappear.[2] The nesting principle, so crucial for Ptolemy, becomes irrelevant for deducing the structure of the physical world.[3] If you insist on the Ptolemaic crystalline spheres in a Copernican world, the spheres have to be puffed up to snugly fit the already known distances, rather than (as in Ptolemy) the minimum feasible size of the spheres determining the distances. So it is not surprising that, although Copernicus himself never challenged (and presumably accepted) the traditional solid spheres, that belief faded as Copernican habits of mind took hold.

With the disappearance of the major epicycles, the retrogressions remain only as an illusion of parallax. An observer (of course) still sees the loops when he plots the paths of the planets. They are really there, but in the sense that an image in a mirror is really there. The loops are still exactly coordinated with the apparent annual motion of the Sun around the Earth—but no longer by some cosmic harmony, for they are all simply images of one annual motion of the Earth around the Sun.

Similar reasoning from the Copernican position of the Earth can be invoked [7.3] to explain why the inner planets stay close to the Sun, why some planets have larger retrogressions than others, why the retrogressions of the outer planets occur only when the Earth is between the planet and the Sun, and so on. Where in the Ptolemaic system the existence of retrogressions and the variations of their timing and scale are empirical discoveries, they are necessary features in a heliocentric world. Further, if particular relations among these features were not present, the Copernican argument must be wrong.

The Copernican and Ptolemaic systems are also topologically incompatible. The Ptolemaic Mercury is always closer to the Earth than is Venus. The Ptolemaic Venus is always closer to the Earth than is the Sun, which in turn is always closer to the Earth than is Mars. But the Copernican Mars, Venus, and Mercury are each sometimes closer to us than the Sun, sometimes farther away, depending on whether the Earth is on the same or opposite side of the Sun from the planet. Mars is sometimes closer to the Earth than is Venus, and sometimes not.

With the move from Ptolemaic to Copernican views, drastic changes also occur with respect to distances. Since the Ptolemaic world must have room for the nonoverlapping spacing of the epicycles, it is larger in terms of distance to the outer planets. The orbits must be spread out to allow room for the epicycles to rotate within their individual spheres. But the absence of any detectable parallax of the fixed stars, despite the orbit of the Earth, means that the Copernican distance to the fixed stars must be vast compared to the Ptolemaic conjecture that the stars lay just beyond the sphere of Saturn. So the Ptolemaic world is about 50% larger than the Copernican in distance from the center to the farthest planet (Saturn); but it is far smaller in terms of distance to the nearest star.

In the Ptolemaic world, the simplest hypothesis to account for the fixed relative positions of the stars during the daily rotation is to suppose that a single sphere carries all the fixed stars together. But on the Copernican view, where it is the Earth that turns while the rest of the universe remains in place, there is no puzzle about how the stars maintain their fixed relative positions. This allows the natural conjecture that stars may vary in brightness because they vary in distance. Copernicus mentions (*De Revolutionibus* I.9) the possibility that the universe could be infinite.[4]

Of course, the most obvious and fundamental point of difference between Ptolemaic and Copernican worlds is that the Ptolemaic Earth is a unique body, fixed motionless at the center of the heavens, while the Copernican Earth is a planet hurtling through space while simultaneously spinning us around its axis at a thousand miles an hour.

10.2

Since the radically incompatible Copernican cosmology undermined 1,400 years of Ptolemaic belief without substantial new evidence, we have an unambiguous illustration of the underdetermination of theory by evidence.[5] If theory (in Leplin's language quoted at the outset of this chapter) indeed is "systematically underdetermined" by intrinsically "unreliable evidence," there might be plenty of room for social construction to govern scientific belief.

But on the whole, the habits-of-mind view goes against that. There are useful senses in which scientific belief can be construed as socially constructed. In chapter 13 I will take that up. But the effects ordinarily (I will argue) fall far short of what is usually implied by constructivist writing.

Sometimes the constructivist claims have been remarkable, such as Harry Collins's (1981) that "the natural world has a small or nonexistent role in the construction of scientific knowledge." But more restrained views have become more typical. Latour invokes microbes and other nonhuman entities as "actants" in an account of social construction, and Pickering talks of "resistances" (by nonhuman entities, such as apparatus) to producing certain sorts of evidence. So the way the world is comes to be treated as an at least implicit factor in social interactions.

From the habits-of-mind view, clarity is better promoted by exploiting, not blurring, a distinction between matters that are significantly subject to social influence and matters that are not.

The point goes beyond the claim that social construction is constrained by a natural world "out there," independent of human beliefs. Insofar as we are capable of knowing anything about the empirical world, that is surely so. If we were given a report that some society had socially constructed a belief that required that newborn babies be purified by boiling, not even the most narrow constructivist would believe it. We all know the world is not made in a way that would allow that sort of belief to survive.

But since we have no logically rigorous way of reaching certain knowledge about how the world is—we can't peek behind the curtain and see what is *really* there—the most we can ever say is that, insofar as human beings are capable of understanding the world, here is how things seem to be. The critical focus then moves from how the world really is from some God's-eye view, to how things seem to be, given the way the human brain reaches whatever understanding it can of such things. Consequently, the focus of the discussion here is on what seems to be in the human head by way of a tuning of cognitive propensities to what works in the world, producing *endemic* propensities [1.1] favoring some beliefs over others. That concern now begins to play a central role in the study.

A philosopher finds it hard to justify an absolute claim that she is not just a brain in a vat, or a program in some hypercomputer. Obviously, she also cannot justify an absolute claim that the world is Copernican, or that the Roman Empire existed, or that she had a grandmother. But outside a philosophy seminar, no one doubts any of these things. Although it is certainly true that, as a matter of logical possibility, theory is always radically underdetermined by evidence, the question for us is how far this portentous assertion has pragmatic significance.

Given cognitive propensities that seem endemic to the species, how wide is the pragmatically plausible span of scientific belief in some context? And how often is sociology more than a peripheral or tactical factor in how such beliefs emerge in the work of some individual and even in how the novelty takes hold (or fails to do so) across a community?[6] Although obviously the contagion, and more subtly the discovery, of a novel idea significantly involves social processes, an important issue—not an issue sensibly kept out of the discussion merely by insisting that everything is social—is whether and in what ways the details of the *particular* social situation actually had any more than a marginal effect on how things went.

10.3

A choice now has to be made. We are interested in the interactions among individual judgment, social construction, and what we can pragmatically suppose we know about the way the world is. But the only practical way to proceed is to set aside some of this for later. In a constructivist account the choice is to focus on sociology. Naturally, I want to start with a focus on individual judgment and the way it is shaped by habits of mind.

So I have been analyzing the rise and fall of Ptolemaic belief, and will be analyzing the Hobbes versus Boyle controversy, in terms of individual judgment. Since I will later argue that taking account of social construction does not in fact add much to these cases (though one occupies a uniquely central position in the emergence of modern science, and the other is the subject of one of the most admired sociological analyses of science), one question that needs to be taken up is whether the very manner of proceeding—setting aside sociology for later—does not prejudge the issue. Naturally, I will try to show that it doesn't.

A second matter of tactics arises, because to conduct any analysis at all requires that we accept some things to start from. There is no way to approach the question of how far the way the world is might shape scientific belief without supposing we know some things about how in fact the world is.[7] This means that the cleanest cases are those far enough back in history that there is no longer any debate about who was right in the controversy at issue. I take it for granted here that (quibbles aside) we know the world is really Copernican, and that (quibbles aside) we know that Boyle was right in his claims about the air pump. This reasonably prompts the constructivist objection that assuming such things as pragmatically certain takes as given exactly what the constructivist analysis wants to make problematical: how did we reach the state where we take for granted some views, and leave their competitors as errors eliminated by some presumptively progressive march of science?

So the discussion of this chapter (and the next) is to be regarded as tentative, contingent on eventually making good on the promise of justifying the tactical choices: treating (in this first cut) individual judgment as something that can usefully be analyzed outside its social context, and treating the successful side of the two controversies that concern us (Copernicus versus Ptolemy, Boyle versus Hobbes) as not just the winning side but the right side.

10.4

If the world is Copernican, then the apparent motions of the planets against the background of fixed stars must be compounded of two independent motions: the planet's own motion, and the heliocentric motion of the moving platform (the Earth) from which observations are being made. So *of course* a reasonably neat account of the apparent motions of a planet is contingent on the combined effects of what I have been calling an s-orbit and a p-orbit [7.3]. There are many observationally equivalent two-orbit models (within the limits of naked-eye astronomy), all connected by the moves introduced in section 7.5. Among

these observationally equivalent systems is the one specifically proposed by Ptolemy. Hence (naked-eye) observations in a Copernican
world must be precisely consistent with the hypothesis that the world
is Ptolemaic. Knowing the world is Copernican lets us understand why
astronomers could believe for a very long time that it is demonstrably
Ptolemaic.

But this correspondence implies some qualification of claims
about underdetermination of theory by evidence. The class of models
that fit the evidence in a sufficiently neat way to stimulate belief among
astronomers was, on the record, extremely restricted. For 2,000 years,
two-orbit models connectable by the three moves of chapter 7 have
completely dominated astronomy. Indeed, as will be discussed a bit
later, only a tiny subset of all workable two-orbit models have been considered. So if theory is radically underdetermined by evidence, as is obviously the case here, the number of theories that the evidence is
underdetermined among is, on the record, tiny, and the theories exhibit
a simple family resemblance in terms of their geometry even though the
cosmologies are radically different. What explains the narrowness of the
set of theories that actually proved capable of winning significant belief
relative to the vastly wider set of logically equivalent setups? Since
whatever constraints are working here were effective over a vast span of
both time and cultures (classical, Islamic, and European), there is a reasonable presumption that what is at work are endemic features of human cognition, not local social conditions. We want to think about what
such endemic features might be.[8]

10.5

Tuning of habits that develop with experience in the world has
to start from something. So it is not at all an unreasonable presumption
that there are some core propensities—what I am calling endemic propensities—that are characteristic of human cognition at all times and
places, genetically entrenched over Darwinian time. Beyond what is in a
strict sense "in our genes," further propensities would develop as genetically entrenched propensities are tuned by experience (with fire,
with falling rocks, with our bodies, and so on) that no human being can
avoid. Endemic propensities may also be entrenched by social processes, but to fit the endemic notion those social processes (involving
discussion, reminiscence, cooperation, and so on) should be so characteristic of human communities that there is no contingent story to be
told about their existence in any particular community.

A presumption that endemic propensity must exist can hardly

be avoided for anyone who accepts a Darwinian view of how to make sense of human cognition. We don't find it a puzzle that the left leg in all the animals we know turns out to be the same size as the right leg. We can leave room for a good deal of nonadaptive variation and still expect that something as grossly maladaptive as wings or legs of unequal size will not survive in a Darwinian world.[9] Similarly, we can expect that cognitive propensities and habits of mind are shaped in ways that ordinarily work well in the world. Many features are common across all human communities, since the entrenched propensities and habits are so generally useful in getting along in the world.[10] I provide a good deal of detail on such matters in the early chapters of *Patterns*.

Behind these propensities is a wider tendency (the "law of effect" [Dennett 1975]) that is endemic across the whole range of life on earth capable of learning at all—which covers essentially all multicelled animals. The law of effect says—or better, simply notices—that ordinarily tendencies that get good results grow stronger; tendencies that get punishing results are inhibited. On the whole, what else could possibly do well in a Darwinian world? On the argument of chapter 1, this should apply to the tendencies of thought peculiarly relevant to human cognition, as well as to the more easily observed tendencies of movement found throughout the animal world.

Further, once a potential belief is on the table, there must be the tension characterized in *Patterns* [2.3] as between jumping-too-soon and hesitating-too-long. Hesitating indefinitely is not a normal possibility. Caught between believing A or B, it is possible you will decide that for now, at least, the question can't be answered and needn't be answered. But a normal person—as Buridan noticed, even a normal ass—will not continue indefinitely in not being able to make a choice when the failure to choose is an impediment to getting on in the world.[11] A person whose ability to act is damaged by a failure to choose will not remain paralyzed between alternatives. So a bystander to a scientific dispute might see the situation as too unclear to make a judgment. But for a participant who needs to get on with her work one way or another, that is almost always cognitively intolerable.[12]

More generally, the endemic human propensity is to believe something, rather than nothing. That implies the propensity to go "beyond the information given," in Bruner's (1957) well-known catchphrase. We believe things that are not demonstrably true. What else could we do, since strictly speaking nothing we believe about the world is demonstrably true? Logical certainty as a condition for belief is not an option for creatures that must get along in the world.

But we cannot resist some beliefs, come to others only with

difficulty, and find ourselves wholly unable to believe yet other things. I want to propose an account of that—oversimplified, of course, but it seems to me useful—that turns on just two points, each of which can be interpreted as a consequence of the law of effect. These will be labelled the propensity to *comfort* and the propensity to *economy*.

A consequence of the law of effect is that we must expect it will be hard to come to believe in a theory that implies X when other things we see do not seem to fit with X. Such a belief will not feel comfortable. But it is tempting to accept a belief that neatly makes sense of things that have caught our interest. Such a belief appeals to our taste for economy. These effects are context dependent, not absolute. Our response to a candidate belief is contingent on how it does compared to the alternatives available. I want to try to pin down a bit what it might mean to say that it is hard to believe something, or easy to believe another thing in some context, and how we choose—or better, how our brains choose, since this is certainly not a matter of conscious will—between rival beliefs. I want to tease out how we might usefully exploit the notions of comfort and economy.

10.6

What makes a belief feel *comfortable* relative to an alternative is that it prompts less sense of conflict with patterns of experience already in the repertoire. A comfortable belief is (relatively) free of rivals, as discussed in detail in *Patterns* (chapter 7). What makes a potential belief feel *economical* is that the additional things you must believe to make this view work seem small relative to the range of inferences you get out. We have a powerful tendency toward beliefs that are economical relative to equally comfortable alternatives, or comfortable relative to equally economical alternatives. The propensities toward comfortable and economical beliefs can be understood as consequences of the law of effect, since the most direct way in which a candidate for belief would be uncomfortable is if our experience is that the idea doesn't work, and the most direct way in which a candidate would be uneconomical is when it is too complicated and narrowly specialized to be of much use. For both, the judgments involved cannot be absolute: the propensities govern our choice among alternatives. Consistent with the law of effect, so long as we see only one possibility that works at all, we make do with it.

These notions of comfort and economy can be defended as just what we see. But it is not hard to provide deeper foundations, since a fundamental and inescapable tension is forced on all creatures that must act in the world by the conflicting risks of jumping-too-soon versus hesitating-too-long. In the language of statistics, we are forced to compromise between the risks of errors of the first kind, acting on a conjecture that turns out to be a mistake, versus errors of the second kind, rejecting a conjecture that turns out to be sound. The propensity to comfort pulls us back from accepting conjectures that do not fit well with our experience in the world; the propensity to economy tempts us to accept a conjecture that looks "too pretty to be wrong."

The clearest cases are where there is no conflict between the two propensities, so that the individual making the judgment finds one view both more economical and more comfortable. For some such cases what seems the more plausible outcome is also obvious to third parties (such as you and me when we come to Hobbes versus Boyle in chapter 11), or clear from other behavior, or both. Indeed, if responses are tuned to the world, then it follows that responses that fit the world particularly well commonly come to be both economical and comfortable. But if a novel belief runs against entrenched habits of mind, that will emerge only over time, the most striking cases being those that have been the special concern of this study, where a belief becomes very clearly both more economical and more comfortable, but only after some *barrier* is broken. Even in less striking cases, it must commonly be the case that economy and comfort evolve with experience with a belief. Economy evolves because of the chunking, or bundling up (with familiarity) of pieces of an argument into a larger pattern that can be apprehended in an all-at-once way, as discussed in *Patterns* [4.6, 5.6, 6.8]. Comfort evolves as habits of mind relevant to the belief are reshaped, which occurs if there is favorable experience with the belief (law of effect), or with generational change as old habits atrophy.

So when an idea is new, there is often tension between the propensity to beliefs that are economical and the propensity to beliefs that are comfortable, even with respect to a belief that will eventually come to be very secure on both counts. Since these propensities are governed by habits of mind, which are not available to introspection and which are never exactly the same for two individuals, it is unrealistic to suppose we could write down agreed formal criteria for economy and comfort, which could then be mechanically applied by some algorithm. But ordinarily we find a good deal of agreement across subjective intuitions, as indeed would be expected among persons responding to similar experience in the world.[13] Where there is sharp disagreement, we can hope to be able to identify why, expecting that this will turn out to be related to conflicts of interest, or varying experience, or both.

Judging which alternative in some particular case has fewer or more marginal difficulties will not always be easy or uncontroversial. But on the argument, we must expect that sometimes the distinction is clear-cut, either immediately or after some barrier of only modest difficulty has been broken. What makes science work, I will argue, turns on its way of seeking to make such choices clear-cut. A good deal of what we are after in these concluding chapters is seeing how that gets done.

10.7

The propensity for economical belief is an extension of a propensity favoring simple beliefs, other things equal. That taste for simplicity reflects the point that, if we live in a reasonably stable world, where similar conditions by and large lead to similar results, creatures would do well to see similar contexts as examples of the same thing. There would be an enormous Darwinian advantage to habitual patterns of intuition that work reasonably well across a wide range of particular situations.

Of course, whatever the prevailing belief, some more complicated belief can be more exactly tuned to a particular situation. But since we will never meet exactly that situation again, that would not be a good way to get along in the world, and creatures with that propensity would not prosper in a Darwinian world. Rather, the simpler the belief (other things equal), the more broadly we will be able to recognize other situations as "looking like" this one.

Restating that argument slightly differently: If the world of experience were chaotic, not reasonably lawlike, Darwinian evolution could not have gotten very far. But if the world must be a reasonably lawlike place, then similar patterns of relations may recur in many different contexts, and beliefs simple enough to look right in many different contexts will tend to prosper, with the provisos that (1) the simpler belief works about as well as its more complicated alternative and (2) the simpler belief does not seriously conflict with other useful intuitions. Simplicity adjusted for the first proviso yields the propensity for economical beliefs, adjusted for the second proviso yields the propensity for comfortable beliefs. If you compare this "simplicity" account of how the propensities for economy and comfort arise with the jumping-too-soon versus hesitating-too-long (errors of the first kind versus errors of the second kind) argument, you will see that these arguments are complementary. Both are ultimately rooted in a Darwinian argument about the law of effect.

10.8

Yet Darwinian stories are not essential for the argument here, since the propensities to economy and comfort are empirically too apparent to doubt, whether you like the Darwinian stories or not. From Aristotle (and no doubt earlier) to our own time, simplicity has been seen as a virtue. The early Copernicans all appealed to it, not as a novel thought but as something whose appeal they could be sure their opponents would concede. Our usual reference is to a version of the principle attached to the name, not of one of the central figures of the Scientific Revolution, but of a Scholastic monk (William of Occam, with his razor of economy). The point to be emphasized is that, whether simplicity (and its elaboration into the tastes for economy and comfort) is explicitly appealed to or not, it is a powerful influence on belief. Its appeal is built into the way our brains work whether we explicitly invoke its appeal in a particular case or not—or for that matter, whether as a matter of explicit principle we endorse favoring beliefs in that way or not.

Similarly, although I give a Darwinian account of why human judgment is forever going "beyond the information given," and Bruner and others have provided broadly similar accounts, that is not essential. A reader with a distaste for Darwinian stories in general, or this one in particular, can look around and see ample evidence that even simple perceptions routinely go "beyond the information given." So a fortiori any theory (that is, any hypothesized account of what lies behind perception) goes beyond the information given. Theory will be extended and elaborated, restrained on one side by the limited imagination of its human users and on the other by getting into trouble if pushed too far beyond what is securely rooted in experience.

Since going beyond the reasons given entails making mistakes, conjectures must sometimes, and bold conjectures often, turn out to be wrong. As all good Popperians would agree—and in fact we are all good enough Popperians for that—a successful strategy of reasonably bold conjecture is contingent on a successful strategy of reasonably severe correction. We can accept bolder conjectures to the extent that good procedures exist for correcting bad conjectures before too much damage has been done.

10.9

What we mean by a "theory" is a story—a pattern of relations—that makes sense of what we think we see directly. It is something abstracted from, or conjectured as lying behind, the surface of direct perception, though (as has been much discussed) what we "directly" see is itself contingent on experience and habit. At one pole are things that we see mediated only by our genetic endowment as that interacts with patterns of experience that are in the repertoire of every human being with normal faculties and experience in the world. Toward the opposite pole are things that would subjectively seem direct (the perceiver is conscious of no intervening steps of inference) only to an adequately prepared individual, as you might directly see a particular painting you have never encountered before as an impressionist painting, though a child accompanying you directly perceives only that it is mostly green.

How can a novel theory emerge? On the account of *Patterns* [9.3], along the path to a discovery at every point the discoverer can be doing only what he already knows how to do. That is, at every point the discoverer must be (can only be) seeing and using patterns that are already in her repertoire, linking those patterns by habits of mind already established. We can allow for some noise in the system, so that what is being seen and done is not *exactly* what is already in the repertoire. But to a good approximation, we see only what we know how to see and do only what we know how to do. New ideas, like new patterns of physical movement, never come ex nihilo.

New theories emerge about what lies beyond what we directly perceive, when we see the context as looking like (roughly fitting) some pattern already in the repertoire. On the argument here, analogy is not one route to discovery, it is the only route (*Patterns,* chapter 6). Since we have a marked tendency to prefer simplicity (not to tune intuition in fine detail to particular things experienced, or to use more entities or assumptions than seem avoidable), a pattern that nicely fits one situation will occasionally be seen as "looking like" another apparently very different situation. Much more often than not, such intuitions prove to be intolerably uncomfortable relative to what we already believe. Occasionally, however, what might be called "fruitful misperceptions" occur.

10.10

In the dawn of science, the pattern that provided the seed for a theory could only be prompted—usually in a wholly tacit way—by things familiar from everyday experience: examples relevant to Ptolemaic astronomy would include experience with how objects behave that exhibit persisting motion, such as spinning tops and wheels, and how such objects look when seen from various angles and distances.

But the propensity to keep pushing beyond the information given must sooner or later go too far. Since we cannot peek in the back of the book to check whether our conjectures are right, our only way of being warned that we have gone too far is to find we have gotten into trouble. Or in the language I have been using: our only way to sense that we have been tempted to be more economical than is prudent is to find that the intuitions that are now being prompted are no longer tolerably comfortable; our only way to sense that we have been tempted to be more comfortable than is prudent is to encounter some alternative, more economical way to account for what we see.[14]

We don't have the propensity to wait until we can be sure for belief to take hold, since a creature with that propensity would forever wait and never act. So it would be naive to imagine that it is a criticism of science or any other arena of human performance that we eventually find we have gone too far. To use statistician's language again, the only way to avoid all errors of the first kind is to guarantee that we will never fail to make errors of the second kind. Effective performance requires a balance between the two. The test of reasonableness is not whether a procedure produces errors, but how effective the balance is between promoting novelty and protecting an ability to retreat when new evidence or argument becomes available to suggest something has gone wrong.

It is not on its face a surprise, or a scandal, or anything but what makes cognitive and Darwinian sense, that sometimes retreat and reformulation occur in science, as in the very radical reformulation from Ptolemaic to Copernican cosmology reviewed earlier in this chapter. But somehow science persistently, though not monotonically, seems to yield a growing core of belief that shows evidence of being a permanent achievement. If that core of "reliable knowledge" (Ziman 1979) is an illusion, we would like to understand how the way things work in the world so powerfully encourages us to see it as right. If it is not an illusion, we would like to understand what accounts for its success.

Here an essential sense of science as socially constructed (in the various senses of chapter 2, note 1) enters the argument, in obvious ways for proponents of the view that science creates an illusion of its own success, but also for the contrary view I will be arguing. Change of belief, as has been discussed at length, will come only when circumstances create a favorable balance for change, which turns on the discomfort and diseconomy encountered with respect to an established belief relative to some alternative, versus the depth of entrenchment and entanglement of the habits of mind that favor the existing belief. But the opportunity for cognitive stress to arise, and the prospects that it will be attended to, is heavily contingent on norms, institutions, technology, and so on, all of which are characteristic of a social context, not of a particular individual.

Anomaly—new information that jars expectations—plays several roles, though at a particular moment one role is ordinarily salient. Anomaly can make an established view less economical than it had seemed, or it can make that view less comfortable than it had seemed, or it can jolt a person from dismissing a transient intuition that ordinarily would have seemed hopelessly wrong, as I have argued in detail for the Copernican case. We have encountered examples of all three situations as this study has unfolded. Typically, more than one of these possibilities will arise in a particular case as discovery unfolds. For as Kuhn has argued, discovery is a process, not an isolated event.

10.11

The gist of the barrier analysis of the Copernican discovery turns on the point that Columbus did not need to look for an unthinkable new continent. He was looking for what he already knew was there—Asia—and he died believing he had reached Asia. Eventually, as exploration proceeded, the radical notion that a new continent had in fact been found on the back of the Earth became too stark to be missed. The Waldseemüller map (*Patterns* [12.8]) lets us date exactly when anyone not closely involved in the exploration—in particular when someone as remote from seagoing activity as Copernicus, landlocked in his "darkest corner of Europe"—might plausibly have become aware that, in stark violation of nested-spheres intuitions, a new continent had been found.

Copernicus published his theory in 1543. But in a passage that refers to the discovery of America, he follows word for word language on the Waldseemüller map of 1507 and the immediately following revision of the world map in a printing of Ptolemy's *Geography* (*Patterns* [12.9]). From other evidence we know that the Copernican discovery—logically ready for any competent astronomer for 1,400 years—in fact came at the very most seven years after Waldseemüller's map (*Patterns* [12.8]). So publication of the Copernican discovery came thirty-odd years after the actual discovery. It took another forty years before we find astronomers ready to follow: a conspicuous case of economy and comfort evolving [10.6] even in the absence of essential new evidence or argument. For soon thereafter, and prior to the evidence of the telescope, Kepler and Galileo both were treating the inverted Copernican scheme of Tycho as the only rival to Copernicus that needed discussion (Margolis 1991).

10.12

In light of the argument to this point, reconsider Leplin's comment [10.1] that evidence is "by nature unreliable" and theory "systematically underdetermined." How well might that claim apply to the radical Ptolemaic-to-Copernican revision of cosmology? Here unreliable evidence was certainly not the critical problem. Copernicus made few observations of his own, and those mainly followed, not preceded, the heliocentric insight. Basically, he relied on what he learned from Ptolemy. Indeed, although it is fashionable to talk of theory-bound observation, which suggests that what is taken for evidence often proves unreliable when theory shifts, in fact it is on the whole easier to think of cases where observation was reliable but ignored—a matter of taming on the account here [3.4]—than of cases where observation was attended to but (in the light of later theory) wrong.[15] Two examples of important tamed anomalies from earlier chapters are the weight gain of calxes up to the time of Lavoisier [4.2] and the failure of Venus to exhibit a variation in brightness, which was never exploited even by Copernicans until the telescope finally made its significance obvious [7.2].

The question of underdetermination of theory needs a more extended treatment than the question of whether pre-Copernican evidence was unreliable. A good deal of the balance of this study is concerned, in one way or another, with the question of how far we can reasonably suppose that endemic cognitive propensities interact with the way the world is, to push theory into channels that are ever more severely constrained, though never completely constrained. Personally, I am not bothered by saying that increasingly constrained theory means learning more and more about how things really are. But defining just what is meant by "how things really are" is tricky and perhaps impossible. For the cognitive issues that concern us, that is not an essential claim anyway. It is enough to ask how far science leads eventually to particular plausible beliefs about such matters as evolution and atoms and the Solar System, as against the possibility that our particular beliefs are only contingent results of how things happen to have unfolded.

In the case of astronomy, economy and comfort, interacting with what astronomers observed, indeed seem to have constrained systems capable of eliciting wide belief to a small subset of possibilities within the very special, and indeed peculiar, class I have labelled [10.4] two-orbit models.[16] Further, once the heliocentric possibility was finally seen, within a few decades *and before any new evidence of consequence had come to hand* (*Patterns* [13.7]) the choice narrowed to a much more constrained subset of two systems (Copernican and Tychonic) in which all nonterrestrial planets, at least, are heliocentric. Within a few more decades everyone knowledgeable enough to make an informed choice and not under church discipline was a Copernican. Economy proved to be a very powerful influence even in this case where comfort certainly initially favored seeing the Earth as stable and central, not as a planet flying through space. As I have argued by now in a variety of ways, perceptions of economy and comfort are substantially contingent on habits of mind, so that, to the extent these are reshaped by new experience and argument, or merely by increasing familiarity and generational change, the balance as it affects a choice between rival theories can drastically shift. Since habits change only over time and are contingent on experience (in particular, an astronomer coming of age around 1580 could not be entrenched in nested-spheres habits of mind in the way that an astronomer forty years older could not escape), the balance between economy and comfort on the Copernican question could shift quite radically over time even in the absence of significant new arguments and evidence.[17]

One could argue that, rationally, the nested-spheres idea that is so crucial for the Ptolemaic cosmology is only a metaphysical gloss on scientific astronomy. Indeed, that is the view (for example) of Price (1959) and Neugebauer (1968). Before the invention of the telescope there was no observation that could tell for or against the nested-spheres idea. What a scientist could measure was only the angular position of various heavenly bodies as a function of time; therefore, what scientific astronomy was about (on this view) was the construction of models that would successfully account for past angular positions and predict future ones. If it was usual to portray the world in the nested-sphere fashion of figure 7.6, that had nothing properly to do with *science.*

Since for naked-eye astronomy, the Ptolemaic, Tychonic, and Copernican systems are all perfectly equivalent [7.5], even if someone believes the world is really Copernican (or whatever else appeals to his taste), when it comes to dealing with observations, he could observe only from his position on Earth, so that perforce the effective set of coordinates would be moved back to the Tychonic setup—which (we've seen) is mathematically just a simple variant of the Ptolemaic. In short (on this view), once you understand the logic of the setup, there is nothing crucial to the choice among the rival systems.

But there is an implausible aspect to such claims. For Neugebauer and Price mathematical astronomy is the figure, the cosmology mere background, where it remained (on this view) until Kepler worked out a version of mathematical astronomy that no longer was just a geometrical variant of Ptolemy. Yet it was Copernican *cosmology* that was important not only to Galileo (who was never much interested in the details of mathematical astronomy) but even to Kepler (who never employed Copernicus's main mathematical innovations). Price credits Copernicus with a great achievement, but immediately insists that Copernicus didn't realize what he was doing. At his conclusion Price calls Copernicus's endorsement of the heliocentric idea a lucky accident. Neugebauer is less generous. Both end up with a portrayal that leaves Copernicus's work as a minor incident, as much an impediment as a contributor, on the way to what eventually became known as the Copernican revolution. I have emphasized the views of Price and Neugebauer, since they are so conspicuously among the very best scholars who have worked on the issue. Price called his article "Contra-Copernicus" and summed up:

> Although Copernicus made a fortunate philosophical guess, his work as a mathematical astronomer was uninspired. From this point of view his book is conservative and a mere re-shuffled version of the *Almagest*. Above all, it introduced many false trails that must have hindered the acceptance of the one good point. In the domain of mathematical astronomy the first major advance after Ptolemy was made, not by Copernicus, but by Kepler. (1959, 198–99)

Neugebauer's (1968, 505) peroration is very similar:

> Modern historians, making ample advantage of hindsight, stress the revolutionary significance of the heliocentric system and the simplifications it had introduced. In fact, the actual computation of planetary positions follows exactly the ancient patterns and the results are the same. . . . Had it not been for Tycho Brahe and Kepler, the Copernican system would have contributed to the perpetuation of the Ptolemaic system in a slightly more complicated form but more pleasing to philosophical minds.

Neugebauer leaves a reader in no doubt that by "philosophical" he means something like "superficial" or "unscientific," so that by the time you are dealing with writers who are following Price or Neugebauer, you can find something close to downright contempt for the idea that Copernicus had done anything really important.

But there must be *something* wrong with such appraisals. Otherwise, we have a very strange coincidence in the fact that among the very few Copernicans before the evidence of the telescope were the two men who came to be recognized then and now as the most brilliant scientists of their period. Further, it was *as Copernicans* (that is, very self-consciously and explicitly so) that these two—Kepler and Galileo, of course—made the discoveries that eventually made a Copernican of everyone else.[18] What each discovered was tied to Copernican cosmology. Galileo's new physics was necessary if sense was to be made of a moving Earth, and Kepler's new astronomy was astronomy guided at every step by the heliocentric way of seeing the world.

In an odd way, a single-mindedly rationalist account can be read to support the otherwise antithetic tendency to deny the usefulness of rationality criteria for understanding science. We might argue—and in a highly qualified way I would argue, and others would do so without much qualification—that the appearance of clear rationality comes into established science only ex post: as cleaning up after the real work has been done and the real battles have been fought. If that is so, Lakatos (1981, 104) was saying more than he intended when he (half-seriously) proposed that the history of science be written as rational reconstructions, with what actually happened (when it doesn't fit such a reconstruction) relegated to the footnotes. From a constructivist view, in particular, that is what most history of science has been doing all along, and without being too careful about the footnotes, either.

10.13

But I want to argue for an alternative to the constructivist propensity to look to interests, tactics, and indeed anything but what scientists claim governs scientific beliefs, *and* an alternative to the rationalist tendency to tell more logical stories than the social and Darwinian characteristics of human cognition warrant.

We get on in the world (I have been arguing) by seeing as making sense what looks like—and only what looks like—a familiar pattern. That is constrained on one side by the patterns available in the individual repertoire, and on the other by the stubborn patterns of experience the world has to offer. In some long-run sense, the latter must in fact shape the former, but since human beings live in a short run, treating the two as independent (for the short run) will work. What we have seen so far is that radically incompatible cosmologies turn out to be based on abstract geometrical models that are mathematical twins. But the rival models are not mathematically equivalent merely by some wild coincidence.

Observations by an earthbound astronomer, to yield an efficient account of the looping features of a planet's motion, *need* two interacting orbits: one to capture the movement around the Sun of the observing astronomer, and another to capture the movement of the observed planet around the Sun. Given the way the world *is*, an economical model can't do with less than two and doesn't need more than two.[19] Further, it is not a surprise that sooner or later someone thinks of the two-orbit scheme. For a person who watches the skies can see (metaphorically, such a person can notice that nature seems to be trying to tell him) that there has to be a double orbit: the planets move around the zodiac, but along the way they occasionally make loops (figure 7.1).

There are only a few ways in which a set of such double-orbit models (one for each planet) could be put together to make a plausibly coherent picture of the world, where plausibility will be contingent on the notions of economy and comfort. What we get is something like a gestalt drawing, which can be seen as making sense in radically different ways, but not in *many* such ways.[20] For Copernicus himself, the possibilities came to be Ptolemaic or Copernican; a few decades later, for Kepler, Tycho, Galileo, and their near contemporaries, the possibilities became Tychonic or Copernican.

That was in a situation where only measurements of angular position were available, and only with naked-eye instruments. Add more constraining details, and the range of essentially distinct, cognitively plausible possibilities must shrink. The range over which theories tolerably consistent with endemic propensities are underdetermined can only grow narrower as observational constraints grow. It is striking—I have tried to emphasize how striking [10.1]—how wide the differences might be between cosmologies that fit exactly the same data. But what we have seen is that at a level intermediate between cosmology and raw data—that of seeing only two-orbit possibilities as credible—Copernican, Tychonic, and Ptolemaic ideas are all part of one closely related family.

Added details sufficient to narrow down that set might not consist of more good evidence, but simply of a good argument that hasn't been noticed before, using no information beyond what has been on hand. Nothing was remotely as important as the sheer force of the Copernican argument in accounting for the moribund state of Ptolemaic cosmology even before Galileo and Kepler made their discoveries. It is uncomfortable to think of the Earth as flying through space at wild speeds, with the fixed stars at some astonishingly remote distance. But on the record, as familiarity with the Copernican alternative grew, it became even more difficult to believe that the planets are governed by complicated motions whose effect is only to make everything look exactly as it would if that machinery were scrapped but the Earth was a planet.

If astronomers appeared to be blind to that possibility even after it was pointed out—as astronomers seemed to be for several decades after Copernicus pointed it out—then we are alerted to look for some entrenched habit of mind that might make that possibility uncomfortable, but in a way that would fade over time as the alternative became familiar. There is no great difficulty discovering what that habit of mind might be, since the well-entrenched nested-spheres sense of the world that lies at the heart of Ptolemaic cosmology, and that (I have argued) provides the most plausible account of what for so many centuries had blocked the Copernican discovery, is also the obvious candidate to account for the forty-year delay after Copernicus's book was published. A habit of mind, like a physical habit, does not disappear once you have a good argument against it. If the way you see the world is guided by the nested-spheres habit of mind, then the marvelous neatness of the Copernican scheme would look hideous. Doing away with the Ptolemaic epicycles does not clean away a lot of unnecessary stuff. For within the nested-spheres view, those epicycles have become essential to the three-dimensional structure of the world [7.3]. Throwing them out would seem like pulling the legs out from under a beautifully set table (*Patterns* [12.3]). But for the next generation of astronomers, who knew of the Copernican alternative from the start, and hence could not be entrenched in the nested-spheres habit before they could consider the alternative, the situation did not look that way at all. Among geocentric astronomers in the 1580s, what stirred a fuss loud enough for us to know about was not a dispute between defenders of Ptolemy and supporters of some version of what we now call Tychonic astronomy, but only rival claims to priority over who was the first to see how to get rid of Ptolemy and still hold on to a stable Earth.

The differences between Copernican and Ptolemaic cosmology are striking. As I tried to show [10.1], a good deal more than whether the Earth orbits the Sun or the reverse is involved. But the eventual convergence of views once the Copernican argument was on the table is also striking. And indeed, if there is a real world out there, then people whose views resonate with what is really out there will have a tremendous advantage over those who have the misfortune to be backing a view that doesn't. The case of Hobbes versus Boyle, taken up next, provides a particularly stark example.

Eleven

Hobbes versus Boyle

Compared to Ptolemy versus Copernicus, the case of Hobbes versus Boyle is extremely simple. But it warrants attention since (through Shapin and Schaffer's *Leviathan and the Air Pump*) it has become a prime example of constructivist analysis. Indeed, precisely the point Shapin and Schaffer stress is that a case that has been treated as open-and-shut by mainstream history of science (to the extent that it has been attended to at all) in constructivist hands yields a rich story of political and social maneuver. Shapin and Schaffer conclude with comments on what they take to be Hobbes's view of the crucial role of politics in establishing the norms of science. Their own summary of the matter then provides a dramatic concluding salvo for the book: "Hobbes was right."

Shapin and Schaffer want to show that there was "nothing self-evident or inevitable" about Boyle's success. Hobbes's views "were not widely credited or believed—but they were *believable;* they were not counted to be correct—but there was nothing inherent in them that prevented a different evaluation" (S&S 13).[1] The habits-of-mind view I am exploring yields the opposite result. Given the endemic cognitive propensities discussed in chapter 10, Boyle's victory was never in doubt, since Hobbes's views were not believable at all. In this case, at least, politics was irrelevant.

On Shapin and Schaffer's account, the effects that Boyle describes in terms of a spring of the air, Hobbes explains as effects of the motions of particles. The particles move through a perfectly fluid ether, the "pure air," which permeates all space, including the interior of Boyle's apparatus. This is certainly a believable argument. Indeed, Shapin and Schaffer's description of Hobbes's position sounds almost exactly like the description that Boyle himself had given of Descartes's view, endorsing it as a perfectly reasonable way to account for what might lay behind the effects he reports.

But as will be seen, while certainly believable, this Cartesian view is not in fact anything like the view actually presented by Hobbes. In a sense, Shapin and Schaffer de facto concede a main point of the habits-of-mind analysis when, in order to make a plausible case for the importance of politics, they slip into attributing to Hobbes a position that is believable, but that Hobbes himself denounced as "incredible" (H 391) and "scarcely that of a sane man" (H 359).

The main points of this chapter turn on why Hobbes's account was not believable in terms of economy and comfort. But to clear the decks, I want to sharpen the contrast between the view Shapin and Schaffer attribute to Hobbes and what Hobbes himself had to say.

11.2

According to Shapin and Schaffer, Hobbes's "fundamental contention" was that

> the atmosphere consisted of a mixture of terrestrial particles . . . and of a pure air. . . . The former fraction mechanically performed the functions which Boyle attributed to the spring [air pressure]; the latter fraction filled the space of the receiver at all times. (S&S 179)

> Again and again, the phenomena that Boyle prized as clear examples of the effects of the spring were appropriated [by Hobbes] as exemplars of the effects of uneven mixtures of earthy and subtle particles moving with simple circular motion. . . . the subtler part of the air was just that part which *rendered a vacuum impossible,* while the grosser parts were those which performed *the effects Boyle interpreted as "spring."* (S&S 122, 123, emphasis in original)

Hobbes indeed begins his discussion (H 358–61) in just the way Shapin and Schaffer describe, climaxing with this very explicit exchange:

> *A* (speaking for Hobbes): "So you have understood my hypothesis: first that many earthy particles are interspersed in the air, to whose nature simple circular motion is congenital; second, the quantity of these particles is greater in the air near the earth than in the air farther from the earth."
>
> *B* (interlocutor): "The hypotheses are by no means absurd. It remains that you show their use." (H 361)

But soon after this promising beginning, Hobbes drops the particles argument and switches to a different and vastly less plausible account. What causes all the effects, Hobbes (1662) says in a summary of the view with which he concludes the attack on Boyle, is a "vehement wind" violently circulating inside the experimental chamber.[2] So there is a Humpty-Dumpty's question here of who is to be master, Hobbes's words or Shapin and Schaffer's interpretation.

11.3

The experiments that led to Boyle's work with the air pump began with the 1643 demonstration that a tube of mercury inverted in a basin of mercury falls to a level about twenty-nine inches above the surface. Torricelli interpreted this as a demonstration that we live at the bottom of what he called an ocean of air. Like fish in the sea, we do not perceive the pressure, since inside and outside our bodies are subject to the same pressure. That the mercury falls, but only to a height of about twenty-nine inches, shows that the accumulated weight of the air at the surface of the earth just balances that produced by a column of mercury of about twenty-nine inches.

If Torricelli was right about the "ocean of air," then it should be possible to get the same effect he had shown with mercury from a tube filled with water, provided that the height of the water column is as much longer than the twenty-nine inches required for mercury, as mercury is denser than water. So if the Torricellian experiment were arranged with a tube of water short of thirty-two feet high (about four hundred inches), water should not fall. But if the column were a little longer than that, then the water should fall, leaving an apparently empty space behind.

When this clumsy experiment was tried (by Pascal in 1647), the predicted results emerged. Many other variants of the original Torricellian experiment were performed, producing many further effects supporting the basic "ocean of air" claim. A particularly simple variant is to tilt the tube so that the top no longer is more than about twenty-nine inches above the surface. As the tube tilts, it immediately refills. So if there is anything in the space above the mercury, it can disappear without much prompting and with no sign of where it has gone.

In a more famous experiment, Pascal arranged for the Torricellian tube to be carried to the top of a mountain. If Torricelli was right, there can be only a limited layer of air above the earth, since otherwise the weight at the bottom of his ocean of air would be indefinitely large. Torricelli had already suggested that air at the bottom of the atmosphere would be compressed to the point where the resistance to further compression just balances the weight of the impinging column of the atmosphere. Pascal—and independently Descartes and Mersenne— therefore expected that atop a mountain the tube would be above an appreciable fraction of the atmosphere. The experiment produced the anticipated fall in what we today would call barometric pressure.[3]

Then in Otto von Guericke's celebrated experiment, two hemispheres were bound together while water was boiled inside to drive the air out. A valve was then closed, the sphere allowed to cool (so the steam would condense), and if Torricelli were right, the external air pressure should exert a large force holding the two hemispheres tightly together. The result was famous; a large team of horses was required to pull the hemispheres apart. Boyle's work with the air pump was intended to provide an observable space larger and more accessible than the top of the tube in the Torricellian experiment, and a space that (unlike Guericke's) would allow observation of what happened to objects inside the space as the air was removed.

All this work dated back into the 1640s, and prompted not only Boyle but Guericke himself, and a group of French investigators that included Hobbes's friend Sorbière (to whom his *Dialogus Physicus* would be addressed) to work on a device that could apply the principle of the water pump to attempt to pump the air from an experimental receptacle. The basic setup is simple to the point of triviality (figure 11.1), but making it work was very tricky, since the piston had to be loose enough so that it could be moved, yet so tightly fitting that leakage remained close to zero even when the pressure difference between the ambient air and the exhausted chamber became large.

Shapin and Schaffer say that "scandalous dissension" within the emerging scientific community prompted Boyle to construct experimental rules of the game that would impose order. But they give no evidence to show that anyone in the seventeenth century thought that the lively

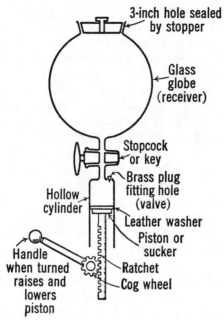

3-inch hole sealed
by stopper

Glass
globe
(receiver)

Stopcock
or key

Brass plug
fitting hole
(valve)

Hollow
cylinder

Leather washer

Piston or
sucker

Handle
when turned
raises and
lowers
piston

Ratchet

Cog wheel

Fig. 11.1. A schematic diagram of Boyle's first air pump. The design was very simple in principle, though difficult to operate with tolerable leakage. With the stopcock open and the valve plugged, the piston is pulled down, so the air that had been in the receiver fills the receiver plus the space available in the cylinder above the piston. Now the stopcock is closed, the valve opened, and the piston pushed back up, forcing out the air that had been sucked from the receiver. The operation is then repeated, with each repetition removing further air from the receiver. Reprinted with permission from J. Conant, *The Overthrow of the Phlogiston Theory* (Harvard University Press, 1950).

debate that Torricelli and Pascal had prompted in the 1640s would be reasonably characterized as "scandalous dissension," or that Boyle or anyone else in the seventeenth century saw what he was doing as laying down rules to prevent such dissension. Nor was there anything peculiarly English about the work on the air pump. What Boyle was doing in England, Guericke was attempting in Germany and Hobbes's friend Sorbière was attempting in France.

Boyle was the most persistent and the most successful of these investigators. He believed he was providing striking new evidence supporting Torricelli's and Pascal's belief that air can be compressed like a spring, or like a ball of lamb's wool, so it takes less space but exerts a greater pressure outward. For example, a lightly inflated bladder should expand (on Boyle's argument) with an increasing pressure difference between the ambient air in the bladder and the partial vacuum created inside the receiver. Indeed, as the pump was operated, the bladder inside the receiver expanded and eventually burst. If pumping was stopped before the bladder burst, and ambient air was readmitted, then the bladder deflated to its original size. Other experiments showed that water at room temperature begins to boil as the pressure falls; sound is muffled; a feather falls directly down rather than fluttering down; and so on.

Overall, Boyle felt he had elaborately confirmed Torricelli's claims about the weight of the air, and had shown many demonstrations of what he called the "spring of the air," the difference between the analogous situation with water pressure being that water cannot be (noticeably) compressed. There is consequently no "spring of the water" to correspond to the spring of the air. Taking a bottle of water up from the bottom of a sea does not yield any effect analogous to Boyle's demonstration with the bladder.

But Hobbes was sure Boyle was wrong about the weight and spring of the air. On his view, since particles in the air seem to just float about, the air in which they are suspended does not gravitate; hence (Hobes argued) air has no weight (H 359). Hobbes was sure Boyle was wrong when he claimed to have weighed a volume of air (H 368–69). So on Hobbes's account the basic mechanism Torricelli, Pascal, and Boyle used to account for air pressure does not exist. Nor can air be compressed. On Hobbes's view, air has a fixed volume, just like water.[4] Hence Boyle's notion of a spring of the air must also be wrong. The spring of the air is, he says, a "fantasy" and a "dream" (H 377). So Hobbes offered a very radical alternative indeed not only to Boyle's views but to those of Torricelli, Descartes, Pascal, and essentially everyone who was a leading figure within the community of natural philosophers. He was very emphatic in denying exactly the two points that Boyle claimed were solidly established: that the air has weight, and that this accounts for what we call air pressure.

11.4

Torricelli had said the space in the tube above the mercury is a vacuum. Many others could not believe that. The Cartesian alternative was to suppose that something more tenuous than ordinary matter fills all space and is capable of penetrating the pores of ordinary material. Since this ether can move right through the glass wall of the Torricellian tube, on Descartes's view, there is no vacuum even if all ordinary air is absent. Boyle says he had no reason to contest such views, since they were compatible with his interpretation of the air pump experiments, and in fact might be right.

Boyle is even more emphatic about the possibility that Descartes might be right in his account of the nature of the air itself, supposing that it is made up of innumerable small particles, with the spring that Boyle attributes to squeezable parcels of air in fact due to the motion of the particles. The particles would increasingly tend to resist further compression as they are driven closer together. Boyle says this might well be right, though he found it simpler for exposition to speak of compressible parcels of air, like balls of wool (Boyle, *New Experiments,* experiment 1).

On Hobbes's side, the key claim is that there can be no empty space in the universe, where he clearly means that atmospheric air (not Descartes's subtle ether) is what completely fills any apparently empty space.[5] This air is perfectly fluid and incompressible. Hobbes judged the commitment to a full universe essential for a proper mechanical account of the world, where all effects are consequences of physical contacts among moving bodies. If all apparently empty space in the world is full of incompressible air, then if you remove a volume from one place (say in the Torricellian tube or in the Boylean receiver), some other volume must fill that space. For the world being full (as Hobbes often says), where else could there be room for it?[6] If Hobbes's view is right, a fit between the piston and the walls of the pump tight enough to allow Boyle to pump out air faster than it could leak back in would be impossible by definition.[7] At all times, on Hobbes's account, the apparently empty portion of the Torricellian tube or of the Boylean receiver is completely full of air. Since the air would have to move very fast to get back in the receiver past the microscopic spaces available around the edges of the piston, it would create a strong wind inside what Boyle claimed to be the evacuated receiver.

In the early pages of his discussion, however, Hobbes does not make the "wind" point, but follows what to us seems a more promising line. Everyone agreed that the air is rich in earthy particles, which can be seen constantly moving about when a shaft of sunlight enters a darkened room. Hobbes argues that the perfectly fluid pure air can squeeze past the descending piston to replace any air Boyle thinks he is pumping out, but the earthy particles cannot do that, and so would be left outside the pump. Hobbes puts these particles to work in explaining various experiments but, as will be seen, *not* in explaining Boyle's experiments. A year after the attack on Boyle, Hobbes published a summary of his views about physics, addressed to Charles II. Here the earthy particles are not mentioned at all. The effects of Boyle's pumping, he says, "is nothing else but a vehement wind, a very vehement wind" (1662, 21), which then accounts for all the results Boyle reports. The 1662 account has only seven pages devoted to the vacuum issue (versus the sixty-one of the 1661 *Dialogus Physicus*), so it is understandable that Hobbes left out what was only a minor part of his argument.

11.5

Since Shapin and Schaffer emphatically state a wholly different position for Hobbes (that the effects that Boyle attributes to the spring, Hobbes accounts for by the motions of earthy particles, so that they attribute to Hobbes the sort of argument that Boyle himself attributes to Descartes), I want to go through Hobbes's argument in enough detail to remove any doubt that, if we allow Hobbes to speak for himself, his position is not at all what Shapin and Schaffer attribute to him.

Hobbes's original involvement in controversy over claims about a vacuum began with his explanation of the Torricellian case fifteen years before the controversy with Boyle. The attack on Boyle itself is most readily understood as an attempt to sustain Hobbes's claim about the Torricellian effect in the face of the greatly elaborated evidence provided by the air pump.

For the Torricellian tube (on Hobbes's account), the space that appears at the top of the tube cannot possibly be empty. The air displaced when the mercury in the tube falls must go somewhere, but since the world is full, the only place the missing volume of air could be is the claimed empty space left at the top of the tube. Although the Torricellian space appears empty, Hobbes says it is certain that it must be full of air. The fall of the mercury, he explains, is to the level where the weight of the mercury in the tube is no longer sufficient to force outside air *through* the mercury back into the tube (H 364).

Hobbes says that pulling back the piston of Boyle's air pump starts the same process as the fall of mercury in the tube. The displaced volume must go somewhere, and (the world being full) the only place it can go is back into the space opening up as the piston is pulled back. So the air can only be forced back into the receiver, squeezing around the edges of the piston. The amount of space for the air to squeeze back into the cylinder past the piston is microscopic. But the amount of air moving into the receiver past the sides of the piston must be equal to the amount being expelled as the piston is pulled back. Therefore the incoming air will be moving very rapidly and create a strong wind inside the space Boyle claims to be empty.

Candles go out and animals die: these do not suffocate for lack of air, as Boyle thinks, but are blown out by the wind (S&S 122). The lid of the receiver, easily lifted before pumping, is now held tight: the vortex holds the lid in, Hobbes explains. The piston of the pump cannot be held down even by a weight of 100 pounds: the vortex causes the air in the receiver to rush out the sides of the piston and forces the piston up, Hobbes explains. The valve of the receiver is opened under water, and water rushes in and fills the receiver. Again, the vortex causes the air in the receiver to rush out, forcing water back into the receiver, though no escaping air is seen. The bladder that inflates and bursts as the pump operates is blown apart by the wind (H 373).

If Hobbes's explanations are correct, it must be the case that air can pass invisibly through water (or through mercury for the Torricellian experiment). For example, when the air pump is operated under water, air being pumped out visibly escapes to the surface. Simultaneously, on Hobbes's argument, the outside air that replaces the air pumped out must be going down invisibly through the water. When the Torricellian tube is tilted, the air flies down the tube and escapes, though that is not visibly what is happening, nor is it easy to imagine how Hobbes would explain why the mercury should refill the tube when it is tilted.

Overall Hobbes dealt with only a few of Boyle's experiments in the 1661 *Dialogus,* and he never explicitly replied to Boyle's response to his attack, or to Boyle's later experiments, for example, showing that a feather falls without fluttering in the exhausted receiver. But he remained very emphatic in asserting that "violent circulation of the air . . . makes all the effects that have appeared in [Boyle's] engine" (Hobbes, *English Works* 7:92).

11.6

So far nothing has been said of Hobbes's use of the earthy particles, which Shapin and Schaffer say provided Hobbes with his alternative to Boyle's claims about the spring of the air. What Hobbes seems to have had in mind are the particles seen in a shaft of light, whose floating quality provided him with his demonstration that air does not gravitate. The effects he attributes to their motions are slight in ordinary conditions, when they are only floating about in ambient air. Here are the effects Hobbes attributes to these particles hitting a surface:

1. Hobbes supplements his main account of the Torricelli experiment with the remark that the earthy particles, impinging on the surface of the mercury, make the Torricellian column fall "a little less" (H 364) than the minimum sufficient to drive air through the mercury. So the effect is explicitly a small correction to his basic argument. Using this point, Hobbes can offer an explanation of why the mercury falls lower at the top than at the foot of a mountain. He argues that the earthy

particles are raised from the surface of the earth by the sun's heat, and that they are less common high above the surface of the earth. Hence at the top of a mountain, the mercury more nearly falls to its minimum level, where the weight of the column just balances the force required to make air pass through the mercury (H 364).[8]

2. Hobbes explains what we call capillary action (the rise of water in a thin tube inserted in the water) by saying that the particles hitting the surface force the water up. As with the effect of the particles on the Torricellian tube, he presents the effect as slight but noticeable because the tube employed is "so very slender" (H 364).

3. Hobbes describes an experiment (actually a variant of an arrangement used by Galileo [1638, 82–83] to weigh air) where water is pumped into a closed vessel filled with air. When a valve is opened, water is forced out. Boyle's explanation would be that the air has been compressed by forcing water into the space. With the air compressed, the spring is increased. So when the valve is opened, water is forced out as the compressed air expands against the weaker pressure of the ambient air. On Hobbes's view air cannot be compressed into a smaller volume. The whole logic of his explanation of the Torricellian and Boylean effects would collapse if that were possible. On his view, as water is forced in, an equal volume of air must be forced back through the incoming water (H 379–80).

Then why is the water expelled when the valve is opened? The motions of the particles provide an explanation. While air can pass invisibly through water (as in his other explanations), the particles that had been in whatever air was forced through the water would be left behind, hence would not have their normal room for circulating motion. Even though his earlier remarks are explicit in treating the pressure exerted by the particles in ambient air as slight, here the particles have been forced together beyond what would be encountered in open air, so that they generate a considerable force.

4. Finally, Hobbes also mistakenly (as he himself later characterizes this claim) invokes the particles in a context that involves Boyle's air pump. Occasionally, the handle controlling the piston slipped from the hand of the operator. The piston would then snap back into a closed position with great force. So the question was, Where does the force required for that come from? On Boyle's account it is the difference in pressure between the inside of the piston, where there is a near vacuum, and the outside, which is under the usual atmospheric pressure. The strong force that had to be overcome to pull apart Guericke's spheres is pushing on one side of the piston, balanced by almost no force on the other. Hence the piston is driven home with great force. Hobbes initial

statement about this uses a variant of the argument just mentioned in 3. As the piston is pulled back (the world being full), air is forced around the edges of the piston into the space that would otherwise be left vacant inside the pump. This is Hobbes's usual point. But the earthy particles cannot squeeze through the spaces left between the piston and its walls. So the particles are left behind, making the region just outside more crowded with particles than is normal. When the hand of the operator slips, the particles regain their normal space for motion by driving the piston home (H 380).

Hobbes changed his mind about that, perhaps because he noticed that while in 3 the particles are confined until there is an opportunity (by opening the valve) to relieve the pressure, here the particles are in the open air. How could more than minimal crowding occur? In any case, in the very last exchange of the *Dialogus Physicus* the question is reopened. The interlocutor of Hobbes's dialogue says he agrees with everything he has been told except the claim about the particles raising the piston. That, he says, is "incredible." What really happens is that the vortex causes the air inside the receiver to rush out around the edges of the piston with sufficient force to drive home the piston. The spokesman for Hobbes then concludes the entire discussion with the remark: "I judge the same. I have erred: and you have rightly corrected my error" (H 391).[9]

Summing up, then: Shapin and Schaffer are simply wrong to say that "again and again" Hobbes explains the phenomena Boyle attributes to the spring of the air by appealing to the motion of particles. In fact Hobbes only claims that this can exert a substantial force in two cases where the particles in some way have been crowded together beyond what would be encountered in ordinary air. He applies such an explanation only once in connection with Boyle's air pump, and that one use he later emphatically disowns as an error, saying it is "incredible" the particles could exert any such force. In discussions of the vacuum after the *Dialogus,* Hobbes never mentions the particles at all. Rather (consistent with the conclusion of the *Dialogus*) Hobbes repeatedly says that all the effects Boyle exhibits with the air pump are to be explained as those of a "strong wind in a narrow room."

11.7

We can now take up Shapin and Schaffer's claim that Hobbes's seventeenth-century readers could have found his argument persuasive had political circumstances been more favorable. If his account were what Shapin and Schaffer say it was (turning on the force generated by the motion of innumerable small particles), then of course it would have been believable. As mentioned at the outset of this chapter, Boyle opens his discussion of the air pump by explicitly labelling Descartes's proposal along these lines as equally plausible with his own description in terms of compressible packets of air. But Hobbes says that explanation is "scarcely that of a sane man" (H 359). Hobbes wanted to rebut Boyle's claim that the spring of the air exists, not give a variant of the Cartesian explanation of it.

Naturally, what is relevant here is the believability of what Hobbes said, not of what Shapin and Schaffer charitably choose to attribute to him. On that basis, the contrast here is as sharp as it could be with the Copernican case, where there was a very apparent conflict between the endemic criteria of economy and comfort. Copernicus (in the relevant sense suggested in connection with chapter 8, note 8) had the more economical theory, but also the less comfortable theory. In that cognitively stressful situation, the popularity for a few decades of the Tychonic compromise is not hard to understand.

But here Boyle's argument wholly dominates Hobbes's with respect to both economy and comfort, so that it is no surprise that, although Shapin and Schaffer insist on the believability of Hobbes's theory, they never mention anyone who actually did believe him.

In terms of economy Boyle's argument is on its face simpler. Hobbes explains most things in terms of the strong wind, but some secondary things also in terms of particles. The Cartesian explanation (which Boyle says at the very beginning of his account is acceptable to him) covers everything by the motion of particles, or in Boyle's alternative explanation all the Torricellian, Pascalian, and Boylean effects are explained in terms of squeezable parcels of air. But the main economy problem for Hobbes is not just a matter of counting entities. It is that the "vehement wind" Hobbes claims is circulating in Boyle's purportedly empty receiver is not an economical idea at all. In application it continually needs to be supplemented by additional claims.

Boyle's argument requires a highly counterintuitive premise: that we live at the bottom of an ocean of air, and spend our whole lives immersed under pressure capable of generating tremendous force. As with the also highly counterintuitive notions of the circulation of our blood and the motion of the earth, we cannot directly perceive the air pressure.

But Boyle's argument comes more than a decade after Torricelli, and in more detail Pascal, had linked the phenomena of the "ocean of air" hypothesis with the more familiar effects of water pressure. We have a view in which the principles that govern the pressure of the air are quite simple and (allowing for the elasticity of the air versus the incompressibility of liquids) yield a general account of fluid pressure that works in both domains. The view readily yields explanations of a very wide range of phenomena unsuspected until this theory prompted the investigations that produced them. The contrast between the diversity of the phenomena explained and the generality and the sparsity of the assumptions needed is just what gives a sense of economy.

Hobbes's argument is very different. For seventeenth-century readers his premise of a full world may have been less counterintuitive than the "ocean of air" idea. Recall that the starting point for Hobbes is that since the world is full, if air is removed from one place, it must be that an equal volume of other air replaces it, since there is no other empty place in the world it could go. The inference that there must be a strong wind in the apparently empty spaces produced in the vacuum experiments is plausible, given the premise. The replacing volume of air clearly must move very rapidly to squeeze around the edges of the piston with no perceptible lapse of time (for Hobbes's claim is that the receiver is always full, not that over time air leaks back in).

But in Hobbes's account there is no link of his explanation to parallel pressure effects familiar from experience with liquids, and (more strikingly) there is no coherent link between the strong wind and the effects he attributes to it. Whatever effect the air pump reveals, Hobbes simply asserts that the wind did it. The vehement wind holds the lid of the receiver in, pushes its way out of the receiver and forces the piston up, muffles sound, makes water boil, blows the life out of animals, blows apart a bladder, and so on: a substantial list of sometimes contradictory effects, whose only common property is that they are whatever Boyle found in the air pump experiments. Instead of getting a lot of explanation out of a few assumptions, we pretty well need a fresh assumption (specifying something else the wind can do) for each result.

And Hobbes's account is very uncomfortable, as well as conspicuously lacking in economy compared to its rival. Everyone has experience with bubbles of air passing through water: the experience is that even a slight leak in a container held underwater produces a readily visible stream of bubbles, and that it takes an easily noticed span of time for those bubbles to move through the water. So it is not comfortable to try to believe, as Hobbes's account requires, that substantial volumes of air pass instantly and invisibly through water. We are all familiar with the way even a slight breeze will disturb a feather. So it is not comfortable to accept a theory that requires belief that Hobbes's strong wind will have no effect at all on a feather left in the receiver.[10]

11.8

Shapin and Schaffer's deeper concern is with why Boyle's style of science prevailed over Hobbes's. They give an account of the contrast between Boyle's position and Hobbes's which makes some sense in terms of what Boyle said Hobbes was saying and of what Hobbes implied Boyle was saying. But Shapin and Schaffer take great liberties in their claim to give an account of what each actually said. Hobbes says that skill in merely doing experiments does not make a man a philosopher, and although he never specifically says so, such remarks are plainly aimed at Boyle. Boyle says Hobbes would discourage doing "nonobvious" experiments, and that if his ideas were taken seriously they would do great harm to experimental philosophy.[11]

But if we turn from how each read the other to what each actually said, the contrast between the two is much less than Shapin and Schaffer assert. Hobbes at no point claims that experimental evidence can be ignored, or that experiments cannot contribute to reasonable judgments between rival theories. Hobbes is very bold in asserting that his principles are the essential foundation of a proper philosophy of nature. But he never claims that a proper philosopher can ignore experimental results, or that his principles must be right whatever experiments might show. On the contrary, the vindication of his principles is that he can (to his own satisfaction, at least) give a better account of Boyle's experiments than Boyle can.

Hobbes in fact seems to balance every sneer at the experimental program with a remark apparently intended to show that his concern is to defend his own views, not to attack the experimental program. The *Dialogus* is addressed to Sorbière, an enthusiastic experimenter who became a member of both the French Academy and the Royal Society. Hobbes introduces the discussion with a remark that "it is not to be doubted that there may be some great consequence for the advancement of the sciences from their [the Royal Society's] meeting. . . . [But] they may meet and confer in study and make as many experiments as they like, yet unless they use my principles, they will advance nothing" (H 347). Shapin and Schaffer were apparently unable to find anything in Hobbes's voluminous writings that puts him in opposition in principle to the Royal Society program, as opposed to insisting that without proper (that is, Hobbesian) physical principles they will not reach a sound interpretation of their experiments. Over and over, Hobbes claims superiority for the Hobbesian interpretation of Boyle's experiments, never challenging the accuracy or relevance of the results. On Hobbes's view, Boyle's experiments, properly interpreted, are like "offerings made intentionally by nature to confirm my physics" (H 347).

Shapin and Schaffer make a great deal of remarks of Hobbes that emphasize that mere skill in doing experiments does not make a proper philosopher. But on this point there is no striking difference between Hobbes's view and Boyle's. Boyle repeatedly makes remarks that, if Shapin and Schaffer's account were right, should have come from Hobbes. The lack of sharp disagreement here is not surprising, since both were expressing the usual view. They differed much less on substance than on Hobbes's propensity to equate sound philosophical judgment with agreeing with Hobbes.

Boyle writes: "I look on the common . . . practices of chymists, almost as I do on the letters of the alphabet, without whose knowledge it is very hard for a man to become a philosopher; and yet that knowledge is very far from sufficient to make one." And "there is a great difference betwixt being able to make experiments and the being able to give a philosophical account of them" (quoted in Partington 1961, 2:498). In an essay instructing a novice in the experimental philosophy, Boyle remarks that he will leave out some mechanical details to protect the interest of tradesmen who market such skills. But there is no loss to the reader in that, he says, since "my aim is to teach you rather philosophy than trades" (Boyle, *Works* 1:315).

11.9

Hobbes never says anything that even suggests he claims that experiments are not central to choosing between theories. Rather, as in the remark quoted earlier, he repeatedly says that his theory gives a *better* account of the experiments than Boyle's does. Hobbes occasionally implies that if his adversaries would adopt his principles they would not need so many experiments: fewer and simpler would do. But he never implies that experiments can be ignored and on several occasions suggests, the more the better.[12] In a discussion explicitly labelled "On the principles and methods of natural philosophy," Hobbes's only comments on experiment advise his reader that "you must furnish yourself with as many experiments . . . as you can," and that what makes a good explanation is that it accounts for the phenomena without "evident contradiction of any manifest truth or experiment" (Hobbes, *English Works* 7:85). The whole rhetorical point of the dialogue form Hobbes uses is that the interlocutor is being persuaded by Hobbes's interpretation of Boyle's experiments that Hobbes's principles give a better account of what is going on than Boyle himself offers. Shapin and Schaffer are driven to quite desperate measures to give the impression that Hobbes is attacking experimental science rather than defending Hobbesian ideas about how to interpret experiments.[13] On the general subject of experiment, Hobbes's rare explicit remarks usually sound very similar to what Boyle would say.

Shapin and Schaffer's claim that Hobbes denies that experiment can provide persuasive evidence for deciding a controversy is fundamental to their wider claim about the political character of making experiment a central feature of scientific practice. If theories can indeed always be quined, controversy would never end. That, Shapin and Schaffer propose, would be dangerous to public order; hence permitting it would violate Hobbes's view of the fundamental responsibility of the sovereign. On Shapin and Schaffer's account, Hobbes's view of the sovereign's prerogative to dictate scientific belief is parallel to his view on the sovereign's prerogative with respect to religion. Shapin and Schaffer cite no statement from Hobbes that plausibly supports that surprising claim.[14]

But if we turn from whether Hobbes actually advocated the views Shapin and Schaffer attribute to him, to a de facto discussion of how he behaved, then Shapin and Schaffer are on stronger ground. Boyle's dedication in contriving and carrying out experiments, caution in speculation that goes much beyond what can be tied to empirical results, and care in arranging that experimental results are validated (by publication of details, repetition, variation, witnessing, and so on) all contrast sharply with Hobbes's style.

Hobbes exhibits a bemusement verging on contempt for Boyle's dogged style, so although it seems to me that Shapin and Schaffer hugely overstate and indeed largely invent Hobbes's principled opposition to experimental science, it is clear that Hobbes's style provides a contrasting model of how natural philosophy might be pursued. The dominant style of science that emerged in the seventeenth century is vastly more like Boyle than like Hobbes.[15] The central claim of Shapin and Schaffer's book is that there was nothing inevitable about that. If social and political advantage had been otherwise, they say, the outcomes could also have gone the other way.[16]

But for both Hobbes and Boyle, their style of doing science was hostage to the substance of their work. Boyle's work was admired throughout Europe. Hobbes aside, even Boyle's occasional critics treated him with conspicuous regard. But highly motivated as they are to show the credibility of Hobbes's analysis, Shapin and Schaffer were apparently unable to find any example of someone who purported to believe, or even to take seriously, Hobbes's account of what the air pump revealed. If no one could be found who was impressed by the fruits of Hobbes's method, how would anyone be attracted to his method? What else should we expect but that the Boylean way with science would be reinforced? Yet even that is not a very interesting conclusion unless you can believe that the role of experiment in the emerging modern science was actually in doubt as late as the 1660s, a generation and more after Galileo, Harvey, Pascal, and Torricelli.

11.10

Since Shapin and Schaffer's book is repeatedly cited as an outstanding example of constructivist work, a question arises about whether constructivist methods are so loose as to permit a scholar to tell any story she wants to tell. Shapin and Schaffer turn inside out what had always been taken as a particularly successful, though not very exciting, episode (since in fact neither Boyle's methods nor his conclusions about the weight and spring of the air are so different from those of Pascal a decade earlier). Shapin and Schaffer turn Hobbes's attack on Boyle into a wonderful story of hidden meanings which set things on a course that has continued to shape our thinking to this day.

But as with Redondi's (humanist rather than constructivist) account of the trial of Galileo [17]—which follows a similar technique of bold glossing of highly selected evidence to reach an astonishing revision of the usual story—are we learning something from this about how things get done in our world? Or are we just being told stories that readers with a taste for that sort of story find congenial?

Twelve

A Note on the Scientific Revolution

It is a very old view that close reasoning pushed up against close observation is the key to how science works. But something more than close reasoning pushed up against close observation is needed to account for the transformation that occurred in the seventeenth century. Harvey (1651), in his methodological introduction to *On Generation,* energetically speaks of what I have been calling close reasoning pushed up against close observation. But he does so in the name of his "leader" and "dictator," Aristotle.[1] So what is new in the seventeenth century?[2]

What I want to point to is a habits-of-mind interpretation of a new and more aggressive sort of experiment, most obviously in the work of Galileo. During the seventeenth century, the means for close reasoning (as well as close observation) was also extended enormously by the increasing application of mathematics in natural philosophy. But as already suggested in the discussion of the emergence of probability (chapter 6), the growing role of mathematics applied to nature was led by the growing aggressiveness of experiment, not the reverse. At least until around 1660, the mathematics used was not very demanding. The ideas are new. But until the work of Huygens and Newton, challenging uses of mathematics occurred only in astronomy (and to a lesser extent in optics), where the tradition dates back to Ptolemy and before. Harvey's work on circulation, then Torricelli's, Pascal's, and Boyle's on air pressure—and even in Pascal's, Fermat's, and Huygens's work on probability and Galileo's work on motion, the mathematics used is not much beyond what could be seen (for example) in the work of the Oxford calculators several centuries earlier. So although the extended—with Huygens and Newton a radically extended—use of mathematics characterizes the emergence of what we recognize as modern science, it is more aggressive use of experiment that plays the leading role. Seventeenth-century usage implies as much. The "new philosophy" was synonymous with "experimental philosophy." "Mathematical philosophy" came only with the *Principia.*

We can see a new and more aggressive style of experiment in place in the work Galileo was doing in the early decades of the century, which (when published) became the focus of intense attention throughout Europe: first through the 1632 *Dialogue,* and then even more emphatically through the *Two New Sciences* of 1638. In the following decade, we see the beginnings of the Royal Society and the great interest across Europe in the experiments on air pressure.

Yet experiment was famously stressed by Francis Bacon in work that precedes any awareness of Galileo's early experimental work, and stressed at least in a programmatic way by Roger Bacon three centuries earlier. Long before that, there are occasional striking examples of what we would unhesitatingly call scientific experiments in the work of Ptolemy and Galen, as will be discussed later in this chapter. But a *novel sense* of what experiment might entail emerged in the seventeenth century. This concerns what Dear (1990) has called "contrived observation," Hacking (1991) has called "artificial phenomena," and (as mentioned in chapter 11) Shapin and Schaffer (1985) and earlier constructivist writers call by such labels as "manufactured facts."

All these writers treat the crucial development as taking place in the second half of the century, putting special emphasis on the appearance of what begins to look like modern canons of reporting of experimental results and of what begins to look like the modern laboratory in the work of Boyle. But the crucial development seems to have been firmly in place a generation earlier. Before midcentury, Torricelli, Guericke, and Pascal had made a generally convincing case for the notion that we live at the bottom of a sea of air. Their evidence was produced by conspicuously artifactual means: even the simple Torricelli tube is a contrivance well removed from ordinary experience—no one would just happen to invert a tube of mercury into a basin of mercury.[3] Guericke's exhausted hemispheres being pulled apart by teams of horses and Pascal's forty-foot tube of wine are even more conspicuously contrived.

12.2

For the emergence of probability, the crucial novelty at the level of habits of mind seems to have been the emergence of a sense that numbers could be constructed, not just counted to, giving us the contrast between direct and inverse counting that is central to chapter 6. A parallel contrast can be noticed between *direct* versus (using Dear's label) *contrived* observation.[4] I want to treat the readiness to look for some way to contrive a context of observation as an emergent habit of mind. It is not, on this argument, an innovation that was consciously designed as a novel tactic, and so something that might have been explicitly and as a matter of principle resisted by people who had another view of the proper methods of natural philosophy. Rather, parallel to the argument about the emergence of inverse counting [6.3], I will argue what had been missing was any routinized (habitual) propensity to look for a way to contrive contexts of observation beyond what direct experience and the focus of attention immediately at hand would provide. Like the sparsely scattered early cases of inverse counting, an occasional instance of what I mean by contrived observation might arise earlier, but only as an isolated departure from what usually happened and what usually continued to happen.

The argument is then that the takeoff of science on a course that yields new ideas at a very rapid pace compared to all previous history turns on the vast increase in the possibilities for application of close reasoning to close observation once the opportunities for close observation relevant to the reasoning are vastly increased by going from direct observation alone to contrived as well as direct observation. This accelerates change and development not only in scientific ideas, but also in economic and technological development, and even in the arts. The wider changes come only rarely by applications of new scientific ideas. Especially in the early decades, what seems most important is contagion of the emerging habit of mind, promoted by the intense interest in science generated by its striking discoveries. Scientists begin to aggressively seek out ways to obtain novel observations, but so do craftsmen and merchants.

On this account contrived observation is solidly in place before—it is taken for granted as background to—the appearance later in the century of modern procedures for constructing, reporting, and validating experiment.[5] The procedural and institutional innovations associated with the founding of modern scientific academies, journals, and so on, are of great interest. But to suppose they are the central development is like supposing that bookkeeping led to the growth of commercial enterprise two centuries earlier, rather than the converse.[6]

12.3

The argument here can be best understood in terms of a parallel with the account given earlier of the emergence of probability, and in a way that interacts with that development. As the emergence of probability seems to be tied to the emergence of a tacit propensity—a habit of mind—that prompts a person to look for possibilities of what (in chapter 6) I call "inverse counting," the parallel here is with the emergence of a more aggressive style of experiment tied to the emergence of a propensity to *look for* a context that could yield observation relevant to a question more immediately in hand. In both cases (inverse counting and contrived observation), we have a shift from taking what is given as immediately relevant to an issue at hand (directly countable things, things directly the focus of the inquiry at hand), to a readiness to look for some way to contrive or discover a context that could turn out to be relevant (something different to count, something different to observe). Here as in the probability case, I want to suggest a barrier that turns on something hitherto missing from the repertoire that enables a critical move, not something already entrenched in the repertoire that blocks a critical move.[7]

This cognitively critical step turns not so much on observation per se, but more particularly on contrived or otherwise indirect *contexts* of observation. Close observation of a situation where some problem or puzzle has come up is endemic. A baby or a cat will qualify as an experimenter if that is all that is meant by experiment. They both readily do some poking and prodding, and watch intently to see what happens. Our instinctive response to a context that catches our attention and is not recognized as something we already know how to respond to is to look closer. With the development of close and extended reasoning— which pragmatically was contingent on the development of writing (*Patterns* [5.13])—we can expect to find evidence of the human propensity to toolmaking applied to facilitate or augment whatever poking or prodding or other sort of observation is going on. Devising tools that can sharpen observation will be stimulated to get poking-and-prodding results exact enough to satisfy close reasoning. The most conspicuous example is that, in essentially all societies where written language has evolved, some form of astronomy has also developed, with contrivances for making and recording observations more precise than unaided vision could provide. As makes sense from this viewpoint, there is often no clear linguistic line—or, as in English, only a recent (postrevolution) line—between words for *experiment* and *experience*. What we are trying to pin down is the emergence of the self-consciously contrived contexts of observation we now peculiarly associate with the label *experiment.*

12.4

We can give a Darwinian account of the shift from direct to contrived-as-well-as-direct observation. Without writing, there could hardly be either the theory and conceptual development that could underlie a search for contrived contexts, nor any means to hold onto whatever might be achieved long enough to make the venture useful. So although elementary experiment—pushing and poking at a focus of attention—probably would become established early in human evolution, contrived observation could hardly be endemic. It could plausibly emerge (like inverse counting) only when favorable conditions had developed. The most salient proximate stimulus here was the challenge to Aristotelian physics posed by Copernican belief. It is consistent with that conjecture that Galileo came to physics (with his early work on motion in the 1590s) as a professor of mathematics, which at that time included teaching astronomy. So he was in a position that gave him the dual advantage of escaping thorough indoctrination in the natural philosophy he would overthrow, and of being well focused on the puzzles raised by the appeal of Copernican reasoning confronted with the physical difficulties of Copernican belief.

Contrived observation is so taken for granted today that it takes some work to see that, until the early seventeenth century, it seems to have been lacking. Deliberately contrived or conspicuously indirect contexts for observation—as distinct from merely contrived means of carrying out observations in a context already the focus of attention—only then became common. The emergence of inverse counting, as I have already suggested, gives us a well-marked illustration of just such an emergent habit of mind [6.3].

What could be taken for contrived or indirect evidence was so rare before the end of the sixteenth century that apparent exceptions are more plausibly seen as the product of special circumstances, parallel to the precedents for inverse counting discussed in chapter 6. There is a famous example from Galen (*On the Natural Faculties,* bk. 1, sec. 13) in which he demonstrates the relation between the kidneys and the bladder; another from Ptolemy, in which he studies the refraction of light; and a scattering of medieval and early modern work by Grosseteste and others, also on optical matters. But Galen (and other work on anatomy both before and after his time) focuses directly on the material of direct interest, though the procedures he uses are certainly contrived, not simple observation. And optics is peculiar in the extent to which the phenomena seen on a scale unmanageably large for poking and prodding present themselves also in familiar tabletop, or at least manageable, scales: we see refractions from a pencil in a glass of water (not just in the distortion of astronomical observations near the horizon), and we can encounter a rainbow in the spray of a waterwheel (not only spanning the sky). Magnetism provides another topic where we can find interesting early examples of experiment, but even more obviously than for optics (until Gilbert's bold extrapolation at the opening of the seventeenth century, discussed next) the work on magnetism focused on material that naturally is encountered at a tabletop scale. Overall, the examples of what look like modern experiments occur so sparsely that each could be—and on the record was—wholly assimilated as familiar, direct observation long before another such example came along.

But at a time when Copernican conflicts with Aristotelian physics provided intense incentive to Copernicans to find ways to discover and demonstrate principles of a non-Aristotelian physics, there was a theory-driven motive to look hard for novel observations, and a theory-driven belief that something might in fact be found. The first published efforts of this sort came, not from Galileo, but from a book that both Galileo and Kepler emphatically admire, and which I would guess was the critical stimulus for Bacon as well. This is Gilbert's *De Magnete,* published at the very start of the century (1600), which reaches striking results from experiments in which a lodestone shaped into a globe is made to play the role of the Earth—Gilbert is most explicit in that: he calls his object a *terrella* (little Earth), and he writes as an enthusiastic Copernican, explicitly so with respect to daily rotation but implicitly with respect to the annual orbit as well.[8] Galileo's work—in the *Dialogue* and much more conspicuously in *Two New Sciences*—is full of artfully contrived contexts of observation. But although published in the 1630s, Galileo's practice of this style of science dates back to roughly the years following Gilbert's book, according to Schmitt's (1969) analysis.

As with the parallel story of inverse counting, there is no sustained opposition in principle to what in practice was a radical innovation. Once a person shows how to attach a number to, say, the weight of the ocean of air, or to probability, for the next person it is no longer cognitively a case of inverse counting. Seeing how to attach the number and, even more fundamental, being prompted to look for a way in which a number might be attached, is a radical step relative to earlier habits of mind. But once that has been done by one person, a second person to whom the result is being shown does not already need that habit of mind to see the sense of what is being proposed. Rather, once Torricelli or Pascal produced the process that yields the number, it took only direct counting to see where the number came from. Hence the tone of the remark of Huygens's English translator, who tells his readers to be prepared to be surprised, but delighted, when Huygens shows them how to attach numbers to chances [6.10]. Similarly, it takes a novel (though no longer so after the first decades of the seventeenth century) turn of mind to be prompted to contrive a context of observation. But once that is done, the audience is just looking at what is the focus of attention. From Aristotle on, natural philosophers had in one way or another routinely invoked the uniformity of nature. Indeed, the very notion of miracles is vacuous unless there is ordinarily uniformity in nature. No one supposed that contriving an observation produced a miracle. Questions could be raised about the relevance of particular contrived contexts to a claim made for them, as such questions continue to arise today in novel situations. But then as now, there seems to have been no objection in principle to contrived as opposed to natural observation. What is new is the propensity to *look for* ways to construct sources of evidence, not the claim that an observation once in hand (from such a construction) might be relevant to another already familiar question. Claims that there were important objections in principle to contrived observation seem to be based far more on twentieth-century speculation than on seventeenth-century evidence.[9]

12.5

There is an element of serendipity here, as indeed is to be expected for *radical* innovation: just what makes an innovation radical is that for some reason the intuition that might lead to a correct conjecture, stimulating a looked-for discovery, is not available. On the argument here, there is a barrier in the form of an entrenched and entangled habit of mind that blocks that intuition, or a habit of mind that is in some way unnatural (something that would not be prompted by endemic experience) needs to develop to make that intuition easily available.

We can see how Gilbert's work could develop out of simple experience with the lodestone, not initially anything like a contrived observation, as direct experiment with lodestones interacted with Gilbert's Copernican interests. Once the lodestone was shaped into and began to be treated as a model of the Earth, we have (with this *terrella*) what seems by a wide measure the most striking example of contrived observation that had yet appeared: a few ounces of magnetized iron shaped into a globe is taken as a model relevant to claims that the Earth rotates. We now would say that Gilbert's explanation of the rotation of the Earth was mistaken. But he was the first to identify the Earth as a magnet and believed he had also shown that rotation is a natural property of a magnet suspended in space.

The practical value of Gilbert's results for navigators was probably not nearly so important for his influence as the mistaken claim that he could give an account of why the Earth rotates. For by this time even non-Copernicans were leaning to the view that the daily rotation at least is carried by the Earth. Tycho was almost ready to concede that. His principle disciple (Longomontanus) did concede it after Tycho's death. By the time Kepler published his *Astronomia Nova* (1609), he seems to treat daily rotation as no longer in controversy. Gilbert's argument—even if we see it as loose and mistaken—could make a great impression on Bacon, Kepler, Galileo, and other key figures of what we now recall as the Scientific Revolution. Gilbert's mistaken claim had the appeal of an argument that found an audience ready to hear it, since they were already comfortable with its conclusion. His argument is about the rotation of the Earth, but what it focuses on is not the Earth, but a lodestone fashioned into the shape of the Earth, providing a most striking example of contrived observation at a moment when the Copernican puzzles made a man like Galileo peculiarly ready to seek more such evidence.

Does this seriously argue that the Scientific Revolution was contingent on Gilbert's error? Or can such instances of serendipity influence only timing, speeding or slowing a development that can plausibly be supposed to be inevitable sooner or later? I want to turn to that sort of question in the concluding chapter.

Thirteen

On Whiggishness

I cannot rigorously prove that I had a grandmother, or that you exist. Since I cannot formally prove even claims that no sane person doubts, I certainly cannot hope to prove a theory, which is about more shadowy stuff than facts, or prove that one theory is better than another. Even an analysis that displays what it takes to be relevant arguments and evidence to explain why some theory displaced another, or ought to displace another, needs to appeal to intuitions about what looks reasonable and to intuitions about what looks relevant, not to formal proofs.[1]

But if empirical belief must always come down to what looks plausible, not to what is logically inescapable, then how could we cut off all the ways in which the play of interests and maneuver and prior commitments might influence what looks relevant and reasonable in a given context? Plainly, on the argument of this study as on other grounds, that could not be done. Beliefs do not come in walled-off compartments, with a section in the mind for religion, another for business, and so on. Hence our beliefs about matters external to science certainly must influence our judgments of what looks plausible within science (and vice versa). Consequently, no clean line can be drawn between internal and external history of science, even allowing (as I would) that a tolerably clean line can be drawn between internal and external factors.[2]

Yet the view of science in this study has been essentially one of rational construction, not of social construction. The accounts I have given mainly concern how barriers were eventually broken. I have tried to show a very large role for habits of mind. But sooner or later, internal factors dominate the accounts. As between Copernicus and Ptolemy, Lavoisier and Priestley, Fermat/Pascal and Roberval/de Mere, Boyle and Hobbes, the former have all been treated as not just the winning side of the argument but the side that on a rational construction of science should have won. There is nothing symmetric in these accounts. Some people get beyond the barrier and get things right, and other people don't and get those things wrong. There is an unmistakable whiggishness here, and since "whiggish" has become almost as unflattering a label as "positivist," a bit of defense is in order.

13.2

Why might science be usefully construed to be a rational construction (in the sense that an extended structure of argument yields compelling inferences about the world beyond direct perception) though the process by which it is constructed is certainly (chapter 2, note 1) one of social construction? Why not? A product needs to be distinguished from the process that produces it, as the intricately refined construction of living creatures that are the product of evolution needs to be distinguished from the blind and chancy process of evolution. The product is the result of the process as that interacts with constraints on what products survive over time. Consequently, the product can exhibit well-marked characteristics that may not be characteristic at all of the process.

In *Patterns* I tried to show in detail why the way individual brains work cannot usefully be characterized as rational, but that human beings by an iterative process can work toward construction of rational (coherent) structures of belief. The social construction of science [10.9] in some ways helps that (since an individual will easily miss his own errors) and in other ways hinders it (since it is so hard to escape what "everybody knows"). But on net, the process of science, on the argument here, works as a cultural development that gradually extends structures of belief into something we can call rational.

We get social knowledge (something for Popper's World 3) that has come to be constructed in a fashion so favored by close reasoning interacting with close observation that it is as compelling to belief as direct perception—and no more in need of qualification by the metaphysical possibility of error (*Patterns* [5.4]). We know that atoms are real and the Earth is a planet with the same pragmatic confidence that we know emeralds will still be green after that gruesome date. The same holds for what is now a vast range of claims that go far beyond habitual induction from direct perceptions. Given that metaphysical certified truth is beyond our reach, what more could be asked of a rational construction of what goes on in the world?

13.3

If you review the earlier chapters, you will find a routinely invoked presumption that today's science provides reliable guidance to what was available for observation at the time of the historical developments examined. This in fact requires no claim (one way or the other) about how far science (in general, or some particular part of science immediately invoked) can be construed to come to better and better approximate what is really there. So the argument is not contingent on any particular view about how far a realistic as against an instrumental interpretation of the science being used is warranted. All that is involved is whether we can reasonably use science that is routinely and reliably used to infer things today about what can be observed by people rolling dice or recording positions of the stars to infer the same for earlier times before this science was in hand—subject only to the obvious qualification that the observations are constrained to means realistically available at the time.

But if contemporary science can be treated *as if* it is right about matters that are not in any dispute, in the cautious sense that things in the world work as if this science is on the whole well tuned to the way the world is, then Tychonic and Stahlian and Hobbesian accounts couldn't have been right. None of these are at all plausible when pushed against evidence no one doubts today.

That does not imply that any particular controversy would have or could have or perhaps even should have been resolved in favor of what today looks like the right side. On the habits-of-mind argument we would not expect that that would always happen, and as a matter of history obviously it doesn't. Only a fraction of the information that now makes the choice—between Copernicus and Tycho, Lavoisier and Stahl, and so on—so completely convincing now was available at the time those choices were live ones. Beliefs—and habits of mind shaping beliefs—that would have been put under great stress had current evidence had been available in the long-ago past could be very secure in their time. We have had occasion to consider why a geocentric version of two-orbit astronomy became virtually unchallengeable though the heliocentric idea had been proposed (by Aristarchus) long before Ptolemy, and why Stahl's phlogiston theory could dominate for decades, though logically evidence that could have prompted an alternative more like Lavoisier's (weight gain of calxes) was well known.[3]

But even in cases where what is being explained is the success of what today seems wrong, the accounts I have suggested are whiggish in the sense that the key to the analysis is an account of why, given the limited information available, or given the force of some barrier habit of mind, what we take to be the better side of the issue did not *yet* succeed. The technique throughout has been to consider what evidence and arguments were available, given the technology available and the logic of the situation, and contrast that with what evidence and arguments were actually noticed and believed. That provided the essential background for considering the role of habits of mind as they interacted with endemic propensities to economy and comfort. From that came accounts of why things worked out as they did. For the habits-of-mind analysis we have been exploring, consequently, a certain whiggishness is built in. Since habits of mind cannot be directly observed [1.6], we can ferret them out only if we can assume that what no one doubts today can be used to tell us what a competent and careful person in historical context could or could not have observed. We need that to detect habits at work from differences between what was seen and what was available, presuming that we in fact know what was available.

But suppose that today's science does *not* provide secure (even if perhaps only instrumental) information about the world, but only belief growing out of a historically contingent victory of one side or another in some now-settled controversy—sometimes the very controversy whose outcome we are trying to explain. We want to consider the plausibility of relativism strong enough to raise this radical indeterminacy as a serious issue: where feedback of theory onto observation can be so strong that we could take seriously the possibility that, had conditions been different before the relevant controversies were settled, we might today be living in a world where Tychonic or Hobbesian or Stahlian claims were believable. A sort of relativism that does not go about this far would really concede the whiggish question.

13.4

Does anyone actually believe in a relativism strong enough to make credible the possibility that, had social conditions, chance events, personalities, and so on been different at the time of the relevant controversies, we might today see a world that looks Stahlian or Tychonic or Hobbesian? Perhaps we should suppose that relativist language is always in fact merely tactical. Or that taking relativism seriously extends only to cases (for example, contemporary cases where issues are still seriously in dispute) where there is nothing surprising about it. On the other hand, it would be presumptuous to insist that extremely able people like Pickering and Collins do not mean what they seem to be intently saying (on Collins's notion of experimenter's regress see appendix B).

So suppose we try to take seriously the possibility that Tychonic rather than Copernican astronomy might have won out, or that Hobbes's interpretation of the air pump experiments might have won out over Boyle's, or more generally that what no one doubts today about the natural world might be very different if historical details at the time the beliefs evolved were different. Does supposing radical social contingency in how science evolves lead to inferences that plausibly fit what we see in the world?

Start here from the most obvious point, which is that science-based technology routinely produces what in earlier times would have been taken to be miracles. We can see what is happening on the other side of the world, hear the voices of people who are dead, travel to the moon, produce in a few seconds computations that once would have required years of labor. Theoretical knowledge developed in the laboratory is routinely taken out of that artificial and protected context and put to use in a thousand unprotected contexts.

Here Latour's interpretation of black-boxing as favoring relativism (since the rest of us become captive to what some center of power has determined is the way to understand the box's performance) seems to point in the wrong direction. The division of labor in any aspect of modern life routinely involves many individuals and many different sorts of skills. In science, as with the division of labor generally, each participant ordinarily takes for granted (accepts as black-boxed) almost all the know-how that gets used in the process. No single individual knows more than a very small fraction of all the knowledge and skills that go into the design, production, maintenance, and operation of an airliner, or a hospital intensive care room, or a particle accelerator. The division of labor allows skills and knowledge to be put to work in places never experienced or foreseen by the originators. But that creates endless opportunities for belief tuned to or determined by the peculiar circumstances of its origination to get into trouble.

Further, what is black-boxed for almost everyone is the special concern of those whose part in the division of labor is precisely to know how to take apart that black box and put together alternatives to it. If we look at a random piece of technology and a random user, black-boxing describes what we will find. But if we look at the division of labor across the social system, and ask if a particular piece of theory or technology is black-boxed for the system, we know that it can't be. There will be people for whom the black box is an open book, and who devote themselves to tinkering with and constructing alternatives to the particular black box that happens to be in use now.[4] Such people take an intense interest in black boxes that behave in ways that violate their expectations, or even that purport to be applications of a theory that in some other, perhaps wholly unrelated, context is discovered to violate experience.

If relativism were right, it is hard to see why science is not continually getting into far more trouble than we see, and very puzzling that scientific disputes overwhelmingly yield consensus, not splitting into antagonistic communities as so commonly occurs in religion and politics. Even science constructed under peculiarly effective conditions of isolation (as with industrial and military rivals) fails to produce the incompatible alternative paths that taking relativism seriously would lead us to expect. Rather, it takes extraordinary political intervention to force that divergence, and even then the results (from the church's commitment to Tychonic astronomy, the Nazis' to Aryan physics, Stalin's to Lysenko's genetics) are short-lived and soon an embarrassment.[5]

The evolution of science is in certain respects like biological evolution in that we see the splitting of lineages into distinct forms (species in biology, specialties within science). But it is completely unlike biological evolution in that we also see the merging of what had been radically different specialties. Splitting within a well-established science is not noticeably into incompatible lineages, but mainly more of the division of labor as particular subspecialties develop within a specialty.[6] And the often spectacular success of mergers (creating molecular biology, or the application of particle physics to cosmology) is astonishing if relativism is right. It seems more reasonable to suppose—certainly no such gross puzzles arise if we suppose—that relativism is just wrong, so that the effects of theory in shaping perceptions of reality are minor relative to the constraining of theory by reality.

13.5

But the argument now has gone beyond what is at issue here, which concerns only whether the fit between today's science and the way things work is good enough to warrant using information that science now provides to inform judgments about what was happening before this information existed, or while it was still in dispute. Against claims to the contrary, we have noticed the effectiveness of scientific theories in applications far outside the context in which they were developed; the lack of fundamental divergence when the opportunity for that is clear, as with military rivals; the successful merger of disparate parts of science in new specialties or new pieces of technology; the frequent synthesis of what had been separate phenomena under some common theory; and (most obvious and fundamental) the persistently successful application of scientific knowledge to do things that were impossible prior to that knowledge. On the other side there is no positive evidence or argument at all, and no explanation of why relativism looks so wrong, if indeed it is not wrong.

13.6

But might there be sufficient tactical advantages to warrant studying science *as if* relativism were right, however hard it may be to suppose that it is? In particular, it might make sense for some people to work from that point of view even if it also makes sense that other people would not. Yet consider some familiar cases. Suppose that we are not allowed to use what we know to help us judge Hobbes's argument against Boyle, Galileo's report of the moons of Jupiter, and Tycho's report that he had measured the parallax of Mars, thereby showing that Ptolemy's system could not be right.

Shapin and Schaffer make a good deal of the point that we do not now have a sample of Boyle's air pump, and hence cannot replicate his experiments. We are denied the opportunity to see for ourselves what Boyle could have seen. Since Hobbes was denied membership in the Royal Society, he also lacked the opportunity to see for himself whether anything went on in these experiments that would have helped his case.

Feyerabend (1975, especially chapter 11) stresses a parallel aspect for the case of Galileo and the telescope. Like Boyle, Galileo enjoyed a privileged position. He told the world what a person should expect to see through the telescope at a time when he had the only effective telescope, and was the only witness to what it could reveal. Some complained, when they had a chance to look, that they did not see what Galileo claimed to see. But Galileo published first, enjoyed by far the widest readership, sent telescopes of his own design to the most influential centers of power, and in general promoted his version of events with rhetoric and action. Doubts about the moons of Jupiter were soon silenced.

Tycho, on the other hand, did not do so well. After his death, Kepler reported that Tycho's claim was a blunder. No such measurement had ever been made, Kepler said, only a calculation of what the parallax of Mars *should* be, given Tycho's belief about the parallax of the Sun and assuming that Copernicus was right in placing the planetary orbits around the Sun.

Consider what can be made of these episodes if we are *not* allowed to exploit what we think we know about what was really there for Boyle or Galileo or Tycho to see. There would then be a good deal of room for argument that Boyle exploited political and social advantages to deny Hobbes opportunities that might have been important to the outcome of the Hobbes-Boyle dispute. That is a persistent theme of Shapin's and Schaffer's book. Similarly, we could put together an argument that Galileo managed to close the door on his critics by techniques that defy canons of fair play and critical reasoning. On Feyerabend's view, Galileo succeeded in evading perfectly legitimate doubts that the telescope could provide reliable evidence. Finally, the possibility arises that it was Tycho's observation, not Copernicus' speculative argument, that was responsible for the collapse of Ptolemy and so for opening the door to a new astronomy. The great prestige that might then have come to Tycho, and thereby perhaps to the Tychonic system, was forestalled by Kepler's claim that Tycho had never made the observation.

This gives us three remarkable claims for the social shaping of scientific belief—two actually made in well-known work, and a third invented for the occasion here, but I think no less plausible. So far as I can see, none of these claims can be sharply answered—or sharply supported, either—from the relativist stance. We are left free to believe whatever we find congenial. But unless we can believe in actual relativism, why should we avoid making use of reliable knowledge (Ziman 1979) that plainly is highly relevant?

If we do use what no one today doubts, none of these claims is plausible at all. If we need to know *exactly* what happened in Boyle's experiments, we would need access to Boyle's actual equipment, restored to its seventeenth-century condition, down to the actual seventeenth-century gunk used to lubricate a piston that had to be tight enough to leak very little even when the pressure difference between inside and outside the pump became large, yet loose enough to be moved. While it would be nice to have that, *exactly* what happened in Boyle's experiments is relevant to the Hobbes-Boyle controversy only insofar as direct observation of the experiments (by Hobbes circa 1660, or by a modern exact replication) might reveal something that could have noticeably undermined Boyle's claims.

But it is only by pretending not to know what no sane person would doubt that that could be an open question. Boyle's reports are consistent with the technology he describes, with the work of Torricelli, Guericke, and Pascal using vacuums obtained in other ways, and also (allowing for improvements in seals and so on) with the pumps that today are routinely used in eliciting Boyle's effects in classes, science museums, and thousands of practical applications. Knowing that what Boyle reports is indeed what he should have found tells us that, as other and better pumps were built, what Boyle reported would be largely replicated in work done by other researchers, using pumps of different design from Boyle's. We are not surprised that within a few years of Boyle's work air pumps were routinely available in the commercial market, as within a few years of Galileo's work the same was true for telescopes.[7]

Free access to Boyle's experiments could not have helped Hobbes, because it is only by an act of tactical denial that we could suppose that Hobbes might then have found significant support for his claim that the effects Boyle reported were those of Hobbes's "fierce wind in a small room." Rather, we know perfectly well (for example) that a feather would lie undisturbed in the evacuated chamber, so that anyone who saw Boyle's experiments or the similar experiments that followed in Paris and elsewhere would find Hobbes's interpretation completely incredible [11.7].

The story is essentially the same for what Galileo reported with his first telescopes. His account generated enormous interest, akin to the interest in the first pictures from outer space in our own time. Within a year, enough people had seen what Galileo had seen through enough different instruments that controversy over the essential reliability of the telescope disappeared. The church's astronomers answered Cardinal Bellarmine's inquiry with assurance that what Galileo reported was there to be seen. As better telescopes became more widely available and experience in how to use the instrument became common, what else could we expect but a strong consensus that what in fact we know was there to be seen could be seen? The tactics Feyerabend uses to build a contrary story illustrate rather than justify his motto of "anything goes."[8]

Finally, we know that Tycho cannot possibly have correctly made the measurements of the parallax of Mars that he reported as a decisive refutation of Ptolemaic astronomy. For we now know that Tycho's parallax of Mars does not fit the world at all. Tycho's estimate is twenty times too large, just as we would expect if what Kepler said were true: Tycho's report is what that parallax should be if Mars were in helio-centric orbit, but the parallax of the Sun were misestimated (as indeed Tycho and earlier astronomers canonically misestimated it) as twenty times larger than it should be. The actual value falls far inside the limits of accuracy of Tycho's instruments. Unless we deny what no one doubts today, we can be certain that the measurement Tycho reports could not have been made.

Nor is the point useful only for checking Kepler's claim. That no one seems to have questioned Tycho's report earlier provides yet another indication that, even before the telescope provided direct evidence that Ptolemy was wrong, belief in Ptolemaic astronomy must have been fading. Otherwise, why was no astronomer of consequence moved to challenge Tycho's report? [9]

13.7

If history of science were a literary genre in which we tell just-so stories to taste, then the relativist stance is exactly what we should want. It maximizes the opportunity to invent a story to suit any taste. But no one who writes history of science claims to be doing that. If we are dealing with what amounts to current events, then a relativist stance may be the only stance realistically available. Too much about what is really there may still be in dispute. And a person might choose, as in Rudwick's much-admired book (1985), to tell a story like a detective tale, holding back how things turned out until the concluding chapters.

In general, for cases where we now know, as well as we know anything, what was really there—for example, that the moons of Jupiter Galileo reported are really there, and Hobbes's fierce wind is not really there—to argue that as a matter of principle we should be satisfied with speculation undisciplined by consistency with what we know is a very strange business. And that holds however pragmatically we interpret "knowing" to mean only that things go *as if* what we call knowledge is true.

Using what we know puts a leash on speculation, but that very discipline also serves as a wedge opening up to reasoned scrutiny things the actors themselves cannot or will not talk about. Judgments about things we have no way to observe directly—about how far religious or political views, or clever rhetoric, or the subconscious effects of habits of mind influenced scientific judgment—can be tested for consistency with claims for which we have special confidence. Our best chance of making persuasive conjectures about what governs persuasion and belief is to anchor what is necessarily a bootstrapping effort in what we know with special confidence.

13.8

It is hard to resist attributing good reasons to actors we have come to admire. If explicit reasons for judgment were given by an actor, then none of us can escape the disposition to accept those reasons as sound if we admire the choice, and the converse in the contrary case. If explicit reasons are not given, we are tempted almost irresistibly to charitably award judgments we like with what we take to be good reasons. For the contrary case, there are the contrary tendencies, attributing the actor's position to some covert interest or bias, and such success as he enjoyed to rhetoric or maneuver or exploitation of misleading habits of mind.

In an older history of science, it was ordinarily an unexamined assumption that scientists who arrived at our present-day beliefs carried the day (when they did) because their arguments were good; if they didn't carry the day, then they were frustrated by the stupidity or prejudice of their adversaries. This sort of writing sooner or later was bound to give whiggishness a bad reputation.

But going by results is not the peculiar characteristic of celebratory history. Whatever the currency an analyst favors, the winners tend to be seen to have had more of it. And of course, being winners, the winning side must have had more of *something.* The problem with what I will call functional whiggishness is that it takes for granted the strength of the winning side along some favored dimensions, then fits (or crams) the story into a frame in which those dimensions are critical.

Of course, this occurs only implicitly: no one would make a transparently circular argument. But to the extent we are left without access to diagnostics that can help discriminate among alternative views of what in fact governed, every writer can be free to read whatever she likes into an episode.

In particular, if there is no useful analysis of the merits of argument, then "anything goes." If you are taken with the role of rhetoric, you can build an account in which rhetoric dominates. Even if there were no use at all of evocative language by the winning side, that would pose no problem. After all, there are contexts in which a homely, restrained style will be effective rhetorically. Rhetoric is like accent: it is always there. In any case, beyond the most trivial cases, advocates use a mixed rhetorical style: there will be arguments framed in a detached, logical way, and others that include poetic or otherwise evocative language. For the winning side, the mix was somehow effective. But it takes some analysis to make an interesting case for how that worked.

A fundamental sociological point here, which is not tied to anything special about science, is that there is a powerful tendency toward synchronization among members of a community. People within a community come to dress alike, talk alike, enjoy the same sorts of music and art, and so on with a virtually unending list of details of life, some obvious, many subtle and ordinarily unnoticed. And that has a functional basis: communities work better when individuals can act with a reliable tacit sense of how others will act, and where reliable interpretation of what people intend by their acts is favored.

The salient illustration is language, which could scarcely work as it does if people within a community did not have an irresistible tendency to come to common intuitions about what words mean. It is an easy Darwinian story to explain why this tendency tends to be general, not limited to a particular set of topics where synchronization is especially important: so we dress alike, as well as understand bits of language alike. This pervasive tendency must have its effects within a scientific community, often generating more consensus on some particular interpretation of evidence than would seem warranted. We cannot make such statements about matters that are merely matters of convention, such as dress or accent. Hence history of science provides a particularly good context for working on the puzzles of social coordination. More than in other contexts, we actually often have a good basis for supposing we know what good beliefs would be in this context. We have the chance to exploit the diagnostic and bootstrapping potential of what science itself can tell us about the world. Although it is merely a convention that we all understand "airplane" to mean one thing and "submarine" another, it is not just a convention that we can actually fly from one continent to another in what we call airplanes. Somehow, there is something right in what we think about how airplanes work.

13.9

Is it naive to suppose there are such things as good arguments? Or are there only what passes for good arguments in some particular social setting? We don't need to do a careful search to know that in no culture is there a belief that castrating all male children at puberty is a good idea. Nature has no trouble giving a clear answer to the question of whether that would be a good idea. But beliefs that are not sharply tested by experience in the world can vary widely across cultures, and they could do so even if in fact there is an important component of what passes for good argument that is shared across wide gaps of time and culture, growing out of what I have characterized as *endemic propensities* [10.5].

Since pragmatic reasoning is shaped by pervasive experience in the world that includes much that is the same for all human beings, what pass for good arguments appear to vary only in limited ways across wide spans of time and place. Biblical arguments that even a creationist would not explicitly make today received careful rebuttals from Kepler and Galileo. Statistical arguments that seem commonsensical today would have been seen as incomprehensible a few centuries ago. Nevertheless, to a striking extent what pass for good arguments are the same everywhere. We read Aristotle because so much of what he said still makes sense to us, and so much of his argument strikes us as shrewd and insightful. As with using what we know about the natural world, using what we know about robustly good arguments can play a crucial diagnostic role. Noticing which good arguments are seen quickly versus which are ignored helps us work out a coherent account of how individual belief evolves, and about the social process that leads to or forestalls the contagion of claims to radically novel knowledge.

The usual relativist response here cites primitive reasoning (commonly, Evans-Pritchard's Azande) as a demonstration that what passes for good argument is socially contingent. The claim is that reasoning in other cultures may be fundamentally different from ours. But the basis of such claims is exceedingly narrow. The same example is used over and over, and it amounts to assuming that, since in one special case (inheritance of witchcraft substance) the Azande seem blind to a *modus ponens* argument, in general they feel comfortable with violations of this most basic axiom of standard logic.

There is no need to look to the Azande for such contradictions. For example, economists in general do not believe that human motivation is strictly self-interested. But nearly all economic theory uses a formal model that allows only for strict self-interest. All competent economic theorists are aware of the contradiction, but by and large they shrug their shoulders and go on, leaving that as a puzzle to be cleared up by someone else or at some other time. Shall we say therefore that economists use Azande logic, in contrast to the rest of us? Nor is it hard to give a cognitive explanation of this commonplace propensity for a community to wall off from ordinarily potent counterarguments occasional incoherence tied to central practices of the community.[10]

13.10

The core issue is not whether science is socially constructed. That is certainly true (in the many senses sketched out in chapter 2, note 1). But it is also certainly true that when I pour sand into a jar, each grain of sand follows its own course, influenced only by local forces. Yet in the end, the sand will be in the shape of the jar, and it would be a very strange and inefficient account of what is going on to try to explain that shape strictly in terms of local forces on grains of sand, ignoring what we know about the jar.

Science, to repeat that most fundamental of points, works by pushing close reasoning up against close observation. Neither could get very far if we could not exploit earlier observations and reasoning that had proven highly reliable, but had to start always from scratch, taking nothing for granted. Popper's (1968) dialectical process of criticism and debate and further testing of what works in the world can refine, by more of the same, what one of us claims to be close reasoning tied to close observation. In principle, nearly all constructivists would agree that acquiring knowledge about how science has acquired knowledge is not a different kind of problem from acquiring knowledge within a science. But relativism would require abstention from the very process that science itself exploits. Here is a place, though not the only place, where taking reflexivity seriously need not merely run in circles.

Appendix A

Introduction to *Patterns, Thinking, and Cognition*

[Bohr] never trusted a purely formal or mathematical argument. "No, no," he would say, "You are not thinking; you are just being logical."
O. R. Frisch, *What Little I Remember,* 1979

This study gives an account of thinking and judgment in which—to lay cards immediately on the table—everything is reduced to pattern-recognition. At least for the tactical purpose of seeing how far we can get with this sort of argument, there is nothing to be said for mincing words, leaving loopholes, making excuses. Further, I will take pattern-recognition as the starting point for the analysis, not as something which is itself subject to analysis. Like atoms for a chemist circa 1900 or genes for a geneticist circa 1950, I treat the central notion as something that is not (here) subject to analysis or even to direct observation. The chemist and geneticist were able to build theories with many observable consequences decades before they could observe an atom or a gene, or say anything at all about what an atom or a gene might really be like. I will try to show that a similarly blunt exploitation of the notion of pattern-recognition can also produce a useful theory.

So I will be working out an analysis of cognition in which everything is reduced to what I will call "*P*-cognition"—to sequences of the

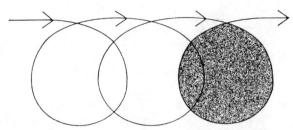

Fig. A.1. P-cognitive spirals. Each spiral represents a cognitive cycle. The arrow at the top of each spiral indicates the prompting of some pattern, contingent on the immediate context, the experience of this individual, and the priming/inhibiting effects of recently prompted patterns. Only a small fraction of these prompted patterns could be expected to come to conscious attention.

sort suggested by the spiraling figure below (see fig. A.1). At the top of each cycle a pattern (the arrangement of features in a room, phonemes in a word, for example) is prompted by cues in the context. That pattern itself then becomes part of the environment which cues the next pattern. Sometimes a pattern, or a feature or subpattern within that pattern, is externally expressed (something is said or done); sometimes (not the same thing) a feature or pattern comes to conscious attention. An externally expressed pattern is more likely to be (or eventually come to be) conscious. Our friends, or enemies, might notice it, even if we don't. Whether conscious or not, though, cues prompt a pattern which, interacting with the rest of the context (hence the spiraling rather than box-arrow-box form of the figure), provides an amended set of cues, which prompts the next pattern. This can go on in several dimensions at once, as when a person plays the piano and carries on a conversation at the same time.

Subjectively (but probably not so at the level of neurophysiology) the patterns are of two contrasting sorts. Sometimes they seem to concern *context* and sometimes they seem to concern *action*. The first is static, like our sense of a familiar face or room. The second involves an unfolding or playing-out, as in a movement or turn of phrase or line of argument. But the distinction between context patterns and action patterns plays little role in the study, and I am skeptical that it has any deep significance. In any case, I will almost never have occasion to explicitly distinguish between context and action patterns, since in context which is relevant is usually evident.

Habits of mind, which play a central role in the argument, should be understood as applying equally to the fluent, automatic prompting of patterns of either sort from the available cues, including among the cues of course whatever patterns (context or action) have been recently prompted.

Further essential complications will develop as the argument unfolds. But the spiral of figure A.1 provides a basic metaphor for the cues-to-patterns-to-cues-to-patterns process.

2

That pattern-recognition is central to thinking is a familiar idea. Everyone who writes on these matters discusses—in some way or another—patterns, frames, schemata, attunements, and so on, which are somehow recognized and then provide a key input to whatever happens next. And I am not alone in making the tuning of patterns of response to patterns of experience not just an aspect of cognition but the central notion (see, for example, Dreyfus and Dreyfus 1986, Lakeoff 1986, Edelman 1985, Rumelhart et al. 1986, and Holland et al. 1986 for diverse ways in which this sense of cognition can be pursued). For the most elaborate computerized character recognizer cannot yet match a bright six-year-old in recognizing letters of the alphabet from many different fonts. No computer, in fact, can yet match a well-trained pigeon in distinguishing pictures that contain some person from pictures that don't (Herrnstein et al. 1976; Abu-Mustafa and Psaltis 1987).

So it is essentially universal to concede an essential role to the cuing of patterns and patterned responses; but the articulation of just what is happening when a pattern is recognized is an unsolved problem. Because so little can be said, the dominant tendency (until very recently, at least) has been to move as quickly as possible from pattern-recognition to some algorithmic, rule-following procedure which can be articulated in a discussion and perhaps instantiated on a computer program. Pylyshyn's recent book (1984, 196–97) (to mention one example of several available) comes very close to claiming that anything that can't be handled in the style of propositional logic—such as learning, affect, and intuition—can be shuffled into a box labeled "functional architecture" and defined to be not part of cognition.

Obviously, I risk erring in the other direction. Unlike the many students of cognition for whom a viable cognitive theory must be implementable as a computer program, I have disowned trying to build a theory of that sort. Pattern-recognition is all there is to cognition, on the account I will give; but no one can say much yet about what the brain is doing when it recognizes a pattern, and I will have essentially nothing to say about that. Then how does the study actually say anything concrete? Without an explicit theory of pattern-recognition, we can hardly expect to deduce what patterns will be perceived in a given context. Yet even without such a theory, given an observation that a certain response has been prompted in a certain context, we may be able to think of (guess, intuit) a pattern that would include that response. We can then suggest that cues appear critical to prompting that pattern. We can then test such a suggestion by observations that alter the cues in some specified way.

There is a good deal of that in the study, increasingly so as we move more directly to applications of the argument in the later chapters. This procedure is applied in a literal way to a sampling of well-known cognitive anomalies (chapter 8), and it is approximated in later chapters in a new analysis of some much-studied aspects of the Copernican revolution. For the historical episodes, of course, we cannot rerun the history with some cues altered as we can replicate the experiments with some cues altered. But the passage of time does something like that for us. The cues and patterns available to Copernicus were different from those available to Aristarchus, Tycho, Kepler, Galileo; and I will try to show how, in terms of the argument of the study, very plausible and perhaps even convincing new accounts emerge of several of the most striking puzzles connected with the Copernican revolution.

3

Another salient question is how the theory handles the empirical fact that human beings, while certainly faulty as step-by-step followers of logical or algorithmic rules, are sometimes able to approximate that sort of performance. We do, after all, follow, and with even more difficulty create, step-by-step sequences like computer programs, mathematical proofs, instructions for assembling bicycles, and so on. But within the argument developed here, that must be done in terms of cues-to-patterns sequences. Rule-following processes, including logic, must be reduced to pattern-recognition, not the reverse. The brain, on the account here, is not properly characterized as illogical or irrational. But it is certainly misleading to call it logical or rational. It is, rather, alogical and arational. What I will try to show is how such a brain can produce behavior and arguments and the theories that are themselves logical, where it is always the product of cognition that is (sometimes) logical, never the process itself.

If cognition turns out to be reducible to computation (and in some sense I suppose that must be possible), on the argument here the steps will not be of a sort we could recognize as looking like the steps in formal logic or in mathematics.

In his *Language of Thought,* Fodor (1975, 28–29) remarks that he takes it for granted that mental operations often have a form that (you can see by looking at his list of properties) looks remarkably like what you will find in the opening pages of a text on decision theory. The brain chooses an act by manipulating sets of possible states of the world and their associated subjective probabilities and utilities. But if the argument here is sound, what goes on in the brain is nothing like that. It would not work that way even in the very special case of a decision-theorist trying to demonstrate how to think that way.

4

P-cognition amounts to a kind of gestalt reversal in the roles of algorithmic (rule-following) processes versus pattern-recognition. Perhaps a way to give some quick insight into the character of this claimed reversal is to point to a feature of the analysis of experiments that is common to leading partisans on both sides of the continuing debate over human rationality but which is not permitted under the argument I will develop. Across the lively debate over "rationality," we find judgment treated (often implicitly, but sometimes explicitly, as in Henle 1962 or in Tversky and Kahneman 1981, to mention two particularly influential papers) as a two-stage process. In the first stage the problem is interpreted (Henle) or framed (Tversky and Kahneman). In the second stage, a judgment is produced, given the interpretation or framing.

Characteristically, the first (framing) stage does not involve an explicit process. Rather, the person is treated as somehow recognizing or imputing a particular way of seeing the problem, in a way that is not itself ordinarily subjected to analysis. Instead, it is only at the second (judgment) stage that we get an analysis, and that comes in terms of a comparison with some normative standard of rational judgment—almost always favorably for Henle and her allies, often unfavorable for Kahneman and Tversky and their allies. For both sides, that second stage is treated as step-by-step articulatable (even when the articultion is not consciously followed by the subjects). For Henle and her allies the process is something close to standard logic; for Kahneman and Tversky and their allies it relies on rules of thumb, heuristics, biases, articulatable not in terms of logic but by procedures that often vary from standard normative accounts (prospect theory, decision-by-aspects).

So one side sees anomalous judgments essentially as consistent with logic, though sometimes logic applied to an interpretation of the problem not intended by the setter of the problem; the other side sees such situations in terms of rule-of-thumb procedures cued by the framing of a situation, often in a context for which it is logically inappropriate (Rumelhart 1981). We then get arguments about whether an anomaly is due to something odd at the framing stage or something odd in the judgment stage. Proponents who see themselves as defenders of human rationality naturally prefer the former. For then it becomes easier to argue that the oddity of framing is only different from, not inferior to, what the experimenter intended; or that it was the impoverished or otherwise peculiar context of an experiment, not anything inherent in the way brains work, that accounts for the anomaly.

In the present study, however, no such division of responses into a framing stage and a judgment stage would arise. In the way that I hope will become increasingly clear as the study develops, we have always the spiraling process introduced at the very start and worked out in detail in the balance of the study. This does not fit neatly on either side of the rationality controversy. In terms of P-cognition, an anomalous response will almost always in fact be a reasonably logical response to another question (as Henle has claimed), and in particular to a question that means something in the life experience of the individual giving the response. But the other question will often turn out to be a logically irrelevant or absurd interpretation of the context that actually prompted the response.

5

There may be no piece of the P-cognitive argument that cannot be found elsewhere. But if there is a fatal flaw ih the study, it is not that it too much merely restates what everyone knows about cognition, or what someone else, or a few other people together, have said. That could not explain the common features reported on such much-discussed material as Wason's four-card test and the Kahneman and Tversky taxi problem. Nor can it explain the new analysis that emerges of so exhaustively a studied piece of history as the Copernican revolution. If I am led to more coherent accounts of these much-discussed experiments and historical episodes, there is a reasonable presumption that I am being helped by a stronger theory than has been available before.

But an unconventional analysis is almost inevitably first seen (above all, first seen by some of its most expert first readers) in a way that lets each bit either look familiar (look like something already known), or confused, or wrong. Hence the pervasiveness of the "where new, not true; where true, not new" response to novelty. Naturally, I cannot realistically feel sure that this study will be among the fortunate ones that survive that phase. But that it will have to go through that phase is probably the most reliable empirical regularity in the sociology of science. As with my previous book, I will follow a distinguished tradition and claim only that the most promising feature of the work is the way that it brings together within a common analysis a wide range of empirical material, which many people would have denied could be so treated.

6

The argument proceeds as follows. Chapter 1 reviews various well-known illusions of judgment and the controversy over their promise as a source of insight into how cognition works. The material is very familiar, and many readers can skim it quickly as stage setting. This is followed by a set of Darwinian arguments which underpin many things later in the study (chapters 2 and 3). While the evolutionary argument includes many familiar ideas, it is more detailed than I have seen elsewhere, hence, I think it is fair to say, more open to both embarrassment and support by empirical work.

In chapter 4, I develop a particular style of talking about the surface phenomena of cognition, in a way that is tied to the evolutionary argument and leading to a more detailed articulation of the notion of *P*-cognition. Chapter 5 then is concerned with the relation between logic and *P*-cognition and with various subsidiary issues tied to that. Chapters 6 and 7 deal with learning: the first with the sort of nonintentional learning that humans share with other animals, the second with the uniquely human sort of learning that involves consciously trying to learn. So I depart from usual practice by allowing consciousness to explicitly enter the argument. I should think that would be a relief to anyone open to William James's remark that it is rather absurd to suppose that the most astonishing thing we know of in the universe is a mere artifact, playing no essential role in how our brains work.

The balance of the study is concerned with empirical applications of the argument. I report some experiments in chapter 8, in connection with new analyses I will be giving of cognitive anomalies demonstrated by the well-known work of P. C. Wason and Kahneman and Tversky. These anomalies will be analyzed in chapter 8 through a purely static analysis, in the sense that we need not consider how changes in the individual's cognitive repertoire might affect the response. In the balance of the study, I extend the analysis to the dynamic context, in which change in the cognitive repertoire is the heart of the matter. For these cases of scientific theory-change we then cannot hope to understand what is happening without taking account of how repertoires—the available patterns and their relations to cues—change (or resist change) over time.

Chapter 9 gives a *P*-cognitive interpretation of Kuhn's paradigm shifts, which are seen as continuous with routine instances of the appearance and spread of novel ideas. Chapter 10 gives some preliminary applications to illustrate the "uphill/consolidation/downhill" schema worked out in chapter 9, mainly in the context of Darwin's discovery. Chapter 11 sketches the background for the Copernican discovery. It starts with a bit of technical detail on the Ptolemaic system that many readers will want to skim quickly and refer to as needed later in the argument. But the balance of the chapter must be read with reasonable care by anyone who wants to understand the new account of the Copernican discovery in chapter 12. Chapter 13 continues the Copernican story into the essentially social context of the contagion of a radical innovation. This crucial stage, I will try to show, was in fact reached around 1590, with the appearance of the Tychonic alternative to Copernicus's system, so that (on this account) the cognitively crucial stage of the Copernican contagion occurs some two decades before the telescope finally revealed the first empirical novelties supporting Copernicus. Finally, chapter 14 makes the transition to fully political judgment by way of a new account (turning on the interaction of political and cognitive factors) of the crisis provoked by Galileo's *Dialog*.

Appendix B

The experimenter's regress argument (Collins 1985) is that successful replication—replication sufficient to account for belief—implies that we know the result of an experiment that qualifies as a good experiment. Otherwise how could we recognize it when we see it? On this argument, widely picked up in constructivist writing, if we know what a good experiment would show, then we already know what the replication is supposed to be convincing us of. If that is so, the question of how we come to be convinced remains unanswered.

But this argument poses a cognitive puzzle only if we treat anything short of strict rigor as worthless. Since empirical belief always turns, and can only turn, on plausible argument, never on formal proofs, it is certainly the case that with respect to replication, as with respect to all other empirical judgments, there is no strict proof. Yet if Collins's argument is put into a context that involves something the reader is familiar with, it is not hard at all to show why replications can be sufficiently secure that no plausible denial remains.

With Galileo's claims for the telescope we have a brand new instrument, wholly unprecedented as a means to study the heavens, applied to give results that for many people were totally implausible and that, even for a Copernican like Galileo himself, were startling. If we could take Feyerabend's analysis seriously, there were ample grounds for resisting Galileo's claims for what he could demonstrate with the aid of the telescope. Further, the episode occurs half a century before the emergence of anything like explicit canons of experimental science, so that convergence on acceptance of Galileo's claims cannot be explained as contingent on entrenched social practices with respect to validation of contrived observations. At the time, no such entrenched social practices existed. If convincing replication occurred, as clearly it did, it must reflect endemic propensities, not some trained, conventional practices of a particular community.

We have in this case the convenience of dealing with a context with which every reader has some firsthand experience. A pair of binoculars is just two small telescopes rigidly linked, so anyone who has used binoculars knows something of the experience of those who replicated Galileo's results, and can draw on that firsthand experience in assessing the situation.

Acceptance of successful replication was in fact very rapid for Galileo. The basic reliability of the telescope and the correctness of Galileo's reports were certainly not seriously in dispute after 1612. Indeed, the initial wave of attacks on Galileo's claims seems to have played itself out within six months or so of his initial publication in 1610 (Van Helden 1988).[1] We have not only convergence, but rapid convergence, and in a context in which all natural philosophers and many astronomers as well were highly motivated to continue the fight as long as opposing Galileo's claims remained a plausible position.

So if there is some special difficulty about resolving conflicting empirical claims where evidence is contrived (a difficulty with experimental evidence that is different from disputes that involve historical evidence or uncontrived observation, and hence warrants a special label like "experimenter's regress"), then we could have expected these difficulties to be conspicuous for this case. Why did the difficulty never seriously arise?

Consider first whether it is true that the only plausible way to decide whether the telescope was performing properly was to compare it with Galileo's report as a standard, hence raising the regress problem. This problem does not arise with respect to objects on earth. We can calibrate a telescope by comparing what we see with it when looking at an object from a distance with what we see when looking at the same object close up. So it is not a surprise that there was never any dispute about whether the telescope worked on objects on earth.

But that would not prove that the instrument must be reliable with respect to objects in the heavens, since only Copernicans believed that heavenly objects were much like, rather than fundamentally different from, sublunary objects. On the other hand, contrary to Feyerabend's insistence, undisputed success on Earth was not irrelevant to the plausibility of success in the heavens. No theory said success was transferable, but also no theory said it wasn't. Once the telescope was available, what had been an obvious impossibility—that we might see what a heavenly body would look like if we could move to within a small fraction of our usual distance—became a possibility: certainly disputable but, also certainly, not obviously wrong. Once many people saw what Galileo saw—which did not take long, due to the intense interest his claims generated—there was no longer room for doubt that what Galileo saw was what other competent observers with a telescope capable of performing well on earthly objects could see in the heavens. And competence can be judged in many ways that do not involve taking what Galileo reported as what a competent observer must see—for example, by competence in using the telescope on local targets. There is no experimenter's regress yet.

Perhaps experimenter's regress arises at the last and crucial stage, where a person must come to believe that there really are moons around Jupiter, mountains on the Moon, and so on, not just artifacts created by the telescope when that instrument is aimed at the heavens. But we all have had experience with sounds at night, things seen in dim light, things recalled from the distant past, and so on, so that we all have some experience with what helps to reassure us that what might be illusory can be believed.

In this case, what was seen through the telescope was different for different objects, so that something unmistakably beyond a common distortion or flaw of the lenses was being seen. If the telescope produced artifactual observations, they were somehow peculiar to particular objects in the heavens. The way the phases and apparent diameter of Venus change relative to its position, the relative speeds and regularity of the moons of Jupiter, and various other effects with respect to the Milky Way and the Moon all fell into sensible patterns that fit the individual objects. If what Galileo was seeing was artifactual, somehow the telescope managed to produce not only totally different artifacts for different targets of observation, but also in a variety of cases to reveal details that make sense in terms of what all parties found plausible: for example, the sixfold variation in apparent diameter that Venus should exhibit whether the real world is Ptolemaic, Tychonic, or Copernican.

So the answer to the question of where validation of the telescope's effects came from is, not from a regress, but from the remarkable coherence of the results. We then easily believe we are seeing something real, as we easily believe a completely filled-in crossword puzzle is essentially correct, without waiting for the solution to be published. In neither case do we have a logical proof we are right. But experience in the world makes it quite impossible to escape the sense that we have gotten things right when many pieces fit together to form a coherent picture: endemic propensities to economy and comfort make that irresistible.[2]

In the event, Galileo quickly won acceptance of astonishing claims, supported only by totally unprecedented procedures, against adversaries who were both numerous and well motivated to contest the reliability of the instrument and the credibility of the observations, given the instrument.

In general, any contrived observation—usually an experiment—produces a signal and some noise. The problem is to discriminate the signal from the noise, and to be confident that the signal is something about the world, not merely an artifact of the particular arrangements of the experiment.

That means that exact replication is not what we want most for novel or disputed phenomena. What was essential for Galileo was that different people, with different telescopes, with the planets in different positions relative to the Earth, reported compatible observations. What was seen exhibited striking coherence, so that (like a completed crossword puzzle) the coherence, if replicable, was powerfully reassuring. Shapin and Schaffer [13.5] were concerned about the lack of exact replication of Boyle's experiments. Yet what a skeptical judge wants to see is whether essentially the same signal—and in particular whether a reasonably coherent set of signals—appears under different but logically equivalent conditions: for if the signal is real, detecting it should be not exactly contingent on one particular occasion or way of observing.

So the characteristic effort within a scientific community seeks to sharpen devices and procedures to a point where the signal, if it indeed reflects something stubbornly out there, stands out unmistakably against the inevitable noise, and not so otherwise. If the signal proves to be robust, we can also expect—and indeed every experimenter hopes—that, as this process goes on, ever subtler phenomena will be brought to light, of which in the air pump case the "anomalous suspension" (discussed in chapter 12, note 7) is an example. It is just the discovery of such novel and unexpected signals about nature (originally, the Torricelli effect; at a later stage, anomalous suspension) that lets science move ahead.

On the other hand, if a reported signal is only an artifact, then there is no reason that it should continue to appear with alternative schemes for detecting it, or that, as the procedure and equipment are refined, it should stand out ever more clearly from the background noise, or that initial suggestions of incoherence should be meliorated rather than aggravated. For a similiar argument in the context of modern physics, see Galison 1988, 257–62.

Notes

Chapter 1

1. For a set of contemporary papers as close to this point of view as any I have run across in mainstream psychology, see Uleman and Bargh 1989.

2. I can recognize a painting by Picasso or a piece of music by Bach, but I can only crudely reproduce it. The pattern in my head is not the pattern in Bach's or Picasso's head, but only some crude approximation of that. And I also lack patterns to control the detailed physical activity needed to produce what I can visualize.

3. How neural patterns are translated into appropriate commands to the muscles is a still largely unsolved puzzle, as is how the pattern of stimulations to the retina are translated into a recognized written word. But those are not problems we need to resolve for the analysis at hand.

4. I try to provide a usefully detailed account of how entrenched intuitions are sometimes successfully challenged in the account of cognitive rivalry in *Patterns,* chapter 7.

5. Branigan (1981) has argued that discovery, on the contrary, is essentially social. But he is using "discovery" in a special way, to mean what would ordinarily be called validation or acceptance of a claim of discovery. That indeed is a social process and of course an important one. It is discussed in *Patterns* under the label *contagion.*

6. In recent decades many discoveries have been the product of a team of individuals working closely together, though I will speak explicitly only of

discovery by some individual. A modestly amended account would also cover the case of team discovery. Even there, in any particular detail of the process, it must be some individual within the team who first sees a point in a way that violates usual habits of mind. On a small scale there will be the usual spectrum, going from what anyone on the team would see or infer in a context, to things that were only seen by a particular team member, though logically others had the opportunity.

7. It is easy enough to think of counterexamples, but not ones that seriously compromise the point. If you and I are both unsure of what to believe, we might settle into compromise beliefs. Or under pressure from an editor or for other tactical reasons a person might be less emphatic or more qualified on a point than he privately feels. Or in a explicitly political context, a person might say things that will help win support. And a person might come to believe some of the things he starts out saying for tactical effect. But it is a fundamental psychological error to talk of negotiating beliefs, as opposed to negotiating claims, statements, and so on. Since believing is not the outcome of a conscious process, or a process subject to direct control of the individual, it is not something a person could negotiate. There is no advantage to conflating negotiation with persuasion, even though it is certainly true that the process of negotiation commonly includes an element of persuasion.

Similarly, Latour and others have argued that scientific controversies are really political, and settled (like other political controversies) by negotiation or coercion. So far as I can see, such claims depend critically on diluting the meaning of the terms to the point where whatever happens falls within the definitions. For example, if you blur the distinction between coercion and persuasion, so that a person who is exposed to the usual argument and comes to believe that the number of primes is infinite has been coerced into doing so, then it becomes tautologically true that all issues of belief are settled by coercion. If there is no good argument, then belief is coerced; and if there is a good argument, then belief is coerced anyway.

Chapter 2

1. There is a straightforward sense in which scientific knowledge is socially constructed, since science could hardly be pursued except under favorable social conditions. To start, since a scientist ordinarily produces nothing he can live from, science can only thrive where institutions exist that allow some individuals to pursue science instead of more directly productive activities. Even given material support, science is rarely a hermit's activity, and never strictly so. Scientists ordinarily require access to teachers at an early stage in their career, access to colleagues at later stages, and socially produced access to information (most obviously, books) at all stages. Although I will argue that, at its core, discovery is intrinsically a story about an individual, the individual story often is not very interesting or instructive (the social context makes the discovery easy enough that the particular individual who makes the discovery is only doing what many others would have done in the same situation). Even in

cases where that individual story is very striking and highly instructive, the individual achievement can only be known to us, and recognized as a discovery, when it becomes a part of established science through social processes. Science itself is intrinsically the product of society, not of any individual within the society, as art is intrinsically a social product, though particular works of art are products of particular individuals.

All these things must influence how ideas evolve and come to spread across a community. What any particular individual takes for granted depends mainly on what "everyone knows." But that a particular individual takes X for granted contributes to everyone else's sense that "everyone knows" X. Fluent communication depends on a broad range of common perceptions of what can be taken for granted versus what needs to be spelled out and perhaps argued about. Powerful effects are at work here even within a community that explicitly cultivates independent and skeptical thinking. Just as we come to share a remarkably subtle common sense of nuance with respect to a shared language, we come to share habits of mind that tacitly but persistently facilitate fluent communication within a community along some lines, but also inhibit communication of novelties that conflict with those lines.

Yet even for language, the common aspect is not completely rigid and universal, or language would not change over time as we know it does. More generally, habits of mind will not be rigidly uniform, for here the particulars of individual experience will yield occasional idiosyncratic departures from prevalent patterns. But core habits of mind that are deeply entrenched and entangled in the practice of the community may be effectively impossible to escape in the absence of some very striking anomaly encountered under exceptionally favorable conditions.

Social considerations enter in another, noncognitive way in the form of influences on what kinds of scientific initiatives will be most generously supported (most obviously, those that have prospects for military or economic payoffs). And occasionally initiatives will be deliberately hindered (most obviously, those that in some way threaten entrenched beliefs in religion or politics). Finally, there are the various further elements of Mertonian sociology of science that shape the allocation of resources within science—the procedures of doing science, the incentives and rewards and sanctions, the norms into which practitioners are trained—all of which are bound to have influences on the emergence and contagion of new ideas.

Most of these elements encourage conformity to usual belief.

2. I treat a candidate for a counterexample (the case of Hobbes versus Boyle) in chapter 11.

As Hull (1988) has argued in a wider and more detailed way, there is no need to suppose that scientists are more devoted to the disinterested pursuit of truth than other people. Features of the activity of science commonly restrain the competing temptations that scientists along with everyone else are vulnerable to. Suppose that I most sincerely wish to avoid eating the wonderful but highly caloric cookie confronting me. In ordinary circumstances I may well find the temptation more than I can resist. But if a madman with an axe is waiting to cut off my foot if I reach my hand out for the cookie, then I am unlikely to be tempted at all. And the scientist who believes—however much philosophers or sociologists may warn against it—that the truth will soon be known, is not much tempted to commit himself to the losing side. Of course, if "interest" is defined loosely enough, we could label any motive (conscious or unconscious, reasonable or perverse) as an interest, so that it is tautological that in some way whatever a person does serves his interests. But that makes the claim that a person is serving his interests vacuous.

3. Kuhn himself has been working on just this issue, formulating a position in terms of changes in language. The account here could be interpreted as complementary to Kuhn's, and it is certainly much influenced by Kuhn's work.

4. I will from here on speak explicitly only of theories, but I mean to include procedures, methodology, equipment, and whatever else is explicitly covered in a textbook.

5. This overstates things a little, but thinking of counterexamples will help reinforce the main point, not undercut it. When Dick Fosbury introduced the "Fosbury flop" to high jumping, there were no doubt—I haven't checked, just because there is *no doubt* this would happen—some who complained that this was no longer *real* high jumping. Therefore although he was winning, it was by a kind of cheating that should be banned. In science, since the point is to find what works in the world, such a response would sound a little crazy. Half a century ago, classical physicists appalled by quantum mechanics said things like, "If this is physics I don't want to be a physicist," not "Even if you get terrific results, this shouldn't be permitted in physics." An interesting, though partial, contrasting response was that of the Brouwer (intuitionist) school of mathematics, who wished to bar from mathematics methods using infinite processes. That was a plausible—but losing—gambit just because mathematics was understood (by mathematicians, if not by outsiders) to be creating beautiful structures of argument, which allowed a role for aesthetic judgment that goes beyond the significant but constrained role that beauty plays in an empirical science.

6. See the comments on a normal-to-revolutionary spectrum in *Patterns* [9.3].

7. Kuhn's notion of revolution is not the only current usage within history of science, and in particular it differs from (hence sometimes conflicts with) the usage of I. B. Cohen in his well-known 1985 study. See Hacking (1986) for comment on the distinction. Obviously, we are concerned here with revolution in Kuhn's sense.

8. I spell out this story in detail in *Patterns* (chapters 11 and 12). As the present study develops, although I will be mainly concerned with new material, the essential points will be recapitulated.

9. Of course *recombinant* DNA did provoke a crisis, but the crisis was almost entirely external (the scientific opposition was essentially confined to individuals active neither in doing nor in managing research.) So very quickly the locus of controversy became legal and political, not scientific.

On the Watson and Crick claims, doubt that DNA was the carrier of genetic information had been based on the apparently simple, repetitive nature of the material: how could it convey information? But the Watson and Crick proposal answered that objection in a most striking way. Technically, their solution applied the results of a fashionable technology (X-ray crystallography) to reach a solution obviously akin to Pauling's widely admired helix structure, and made immediate sense of a hitherto unexplained regularity of DNA across species (the Chargaff ratios). If there had been a marked delay in reaching this discovery, or marked signs of incommensurability once it was on the table, we would expect to find a habits-of-mind explanation of that. But, as will be stressed in chapter 3, it takes unusual circumstances for such a barrier to be present. But we could expect to find some milder indications of incommensurability even in this case.

Chapter 3

1. Thagard (1990) provides a carefully worked out gap account of scientific revolutions as remappings of networks of explicit rules, taxonomies, and so on, applied in particular to the overthrow of phlogiston. The account can be compared to the very different one I will give here of the same episode.

2. A constructivist might object to the implicit claim here that arguments are "strong" or not so in some unproblematic way, as opposed to that perception of strength being shaped by the social context. On the view I am trying to develop here, it is indeed meaningful to talk about an objective basis for judging the strength of a case: one that works imperfectly but still impressively and usefully across wide variations in culture and context. I develop that argument in some detail starting in chapter 10.

3. What about overt evidence of incommensurability? That can be set aside in various ways. The indications might seem ambiguous, or not strong enough. Or talk that looks like an inability to see the point of a new argument might be written off as merely tactical or rhetorical, or (less fashionable today) as merely reflecting the stupidity or bigotry of the resisters. Most of all, since acceptance of new empirical ideas is always a matter of judgment, it is always possible to argue that resistance to the idea was perfectly reasonable, on the logic of the situation. Adversaries ordinarily provide some arguments for their view—human beings are almost never at a loss to give some sort of argument to explain their positions. And if the adversaries fail to provide an argument, or provide only arguments that look foolish to today's readers, it is always possible to suggest arguments to justify resistence. So one way or another, the opportunity to make a claim that resistance was rational (for positivists) or reasonable (considering the social context, for constructivists) is always present.

4. The most important counterexample I have run across is Butterfield's (1965, 29) suggestive (not fully explicit) remark, more fully quoted in *Patterns* [12.6]: "Copernicus . . . lost . . . the whole intricately dovetailed system in which the nobility of the various elements and the hierarchical arrangement of these had been so beautifully interlocked."

5. Naturally success words like 'discover,' 'know,' 'recognize,' and 'understand' here have to be read as qualified by phrases like 'looks like,' 'seems

to,' or 'thinks he.' Similarly, a distinction is required between my subjective sense that I have discovered or know something, and the social process of validation, or contagion, of the ideas, which (in favorable cases) yields a sense shared across the community that "everyone knows" that what I think is a discovery in fact is a discovery. See the comment on Branigan (1981) earlier.

6. Two subsidiary arguments for uniqueness are, first, the conditions needed for a habit of mind to function as a stubborn barrier are unusual. So the first point that suggests that some particular barrier might be unique to each such case is just that we do not expect exceptional things to occur very often. Second, as I have already mentioned but not yet illustrated, the habits of mind we find when a concrete case is examined from the barrier viewpoint turn out to be so directly linked to the critical intuitions that there could hardly be more than a very few plausible competitors for the role of barrier.

But neither of these arguments is strong. It must indeed be rare for multiple, well-entrenched, and entangled potential barriers to be present. However, it is also rare that marked symptoms of incommensurability characterize a discovery. It may be that a fairly *common* condition among these rare episodes is the presence of more than one potential barrier, making the breakthrough effects available from getting past the first barrier less stark, hence more easily missed during the limited window of opportunity.

Chapter 4

1. See, for example, Donovan 1988 and its companion articles, which are largely devoted to this theme.

2. Both Lavoisier and Priestley at first failed to realize that the gas released when intensely heated oxide of mercury resumed its metallic form was something different from atmospheric air. Lavoisier got things clear only after Priestley had discovered his own error and publicly corrected Lavoisier's. Conant (1950, still the most useful of the "usual accounts") gives a very clear presentation of this and of many other details.

Scheele was the first to get the whole story, but the last to have his account appear in print. All three had the discovery in hand by the end of 1776. But none immediately saw oxygen as making phlogiston redundant, and two of the three never accepted that. With Priestley, it is also hard to say for certain whether he saw dephlogisticated air as a distinct component of the atmosphere, or only as atmospheric air with less than the usual loading of phlogiston (with nitrogen the same but saturated with phlogiston). On this as on many other points, Priestley is very hard to pin down. Late in life he said he couldn't be accused of holding out for phlogiston out of mere stubbornness, since in fact he frequently changed his mind, as indeed he did.

3. Since the metal was produced from ore by heating with charcoal, there was commonly some contamination with carbon, hence detectable carbon dioxide was commonly produced. But standard techniques were available for washing out this component. The point here is that there was no phlogisticated air (nitrogen) produced. In general, contamination of materials and other

sources of misinterpretation complicated the story beyond what I can detail here. Kitcher (1993) gives an account that stresses this aspect of the issue, and there is no doubt it is important. But conflicting evidence is the usual, not an unusual, situation where novel effects are at issue. On the barrier view, that is important background information, but not ordinarily what differentiates the easy predominance of some new ideas and the stiff resistance to others.

4. For Musgrave, Lavoisier's success reflects the correctness of Lakatos's methodology. He argues that by the mid-1780s the Stahlian theory had been subjected to so many ad hoc revisions that it was recognized as a degenerating research program. But "the 1784 phlogiston theory had as much evidential support as 1784 oxygen theory" (Musgrave 1976, 205). I will argue otherwise in chapter 5.

5. The usual view divides into two schools: an older school that treats Lavoisier as a revolutionary hero, and a more recent one that treats what he did as so contaminated by the obvious and the erroneous (his theory of acids, his theory of heat) that the puzzle is why Lavoisier's overthrow of phlogiston should be called revolutionary. This revisionist view is peculiar both in pushing phlogiston out of its position as the keystone of Lavoisier's work, though both Lavoisier and his adversaries explicitly thought it was, and in emphasizing other points, on which in fact there was no sharp division between Lavoisier and his adversaries.

The development of chemistry was still at a stage where the notion of a chemical bond had yet to emerge (hence the distinction between a compound and a mixture was not yet clear). The notion of conversion of energy (as from mechanical to electrical) also lay in the future. The partiality for the view that heat is a fluid (versus the view that heat is the motion of atoms) was based essentially on the difficulty of accounting for the phenomenon of latent heat if heat were motion. Today we would say that the kinetic energy of perceptible heat is transformed into the electrical energy of the chemical bond. But at a time when electricity itself was thought to be a fluid, and chemical bonds and the transformation of energy were still unknown, it was easier to think of heat as a fluid. That fluid might be absorbed up to a certain point by materials, so that only heat beyond that point of saturation would be perceptible as an increase in temperature. The main advocate (circa 1800) of the mechanical view was Count Rumford, whose supervision of the boring of cannon barrels gave him intense experience with an operation that favored the mechanical interpretation of heat.

The prevalent preference for thinking of heat as a fluid is wholly irrelevant to the controversy over phlogiston. Lavoisier took the usual position here, favoring the fluid view but allowing that the mechanical view of heat might ultimately prove right. He eventually, as part of the general reform of chemical nomenclature, provided the name *caloric* for heat as a fluid. But there was no theory of heat peculiar to Lavoisier. Similarly, the belief that oxygen was a necessary component of acids, though in this case originating with Lavoisier, was accepted by Priestley, Scheele, and other defenders of phlogiston.

6. A common claim is that Lavoisier's theory surrendered the ability of the phlogiston theory to explain the special properties of metals: metals were

shiny, malleable, and so on, due to the presence of phlogiston. However, I have seen no evidence that this was an important argument for Lavoisier's contemporaries, nor is it very plausible. After all, many things taken to contain phlogiston (such as sulfur or charcoal) were not metals. And only a very few "earths" could be heated with charcoal (thus adding phlogiston) to produce a metal. Apparently the substances for which this worked (the calxes) already shared some property that made them suitable for this transformation. Of course if the existence of phlogiston was *not* at issue (it was taken as uncontroversially established), then it would be tempting to draw this known constituent of metals into an account of the properties of metals. But if the *existence* of phlogiston was the issue, the tenuous claim that it could account for the properties of metals is scarcely an argument worth making. On the analysis to follow here, there is no reason to lean on this weak reed to help account for the resistance to Lavoisier's claims.

7. Lavoisier is commonly quoted in a way that gives the contrary impression, by focusing on one phrase: his argument would "shake Stahl's theory at its foundations." But Lavoisier's paper qualified this stark-seeming claim at a number of points, including within the same passage allowing that his theory and Stahl's explain things equally well. The paper as a whole does not at all sound so bold as the phrase "shaken at its foundations" taken out of context would suggest. And indeed even that phrase is contingent on an "if" clause which is routinely left out of the quoted language.

8. From Lavoisier's 1777 paper (translated in Conant 1950): "my object is not to substitute a rigorously demonstrated theory [for Stahl's], but solely a hypothesis which appears to me more probable, more conformable to the laws of nature, and which appears to me to contain fewer forced explanations and fewer contradictions."

The analogous passage from his post-1783 "Reflections on Phlogiston" says his object is "to show that the phlogiston of Stahl is an imaginary thing whose existence has been gratuitously supposed. . . . all the phenomena of combustion and calcification may be explained in a far simpler and easier manner without phlogiston than with it" (translated in Musgrave, 1976, 203).

9. Perrin's criteria (1988, 115) are looser than necessary to closely test the argument here, where the crucial issue is denial of the reality of phlogiston. He lists Montigny and Bayen as converts prior to 1777 (that is, prior to Lavoisier's own espousal of that position). He lists Busquet and Laplace for 1777–78, but apparently mainly on the grounds that, as close junior collaborators with Lavoisier, they showed themselves to be in sympathy with his views. No one else is listed until 1784.

10. Lavoisier was the first to see airs as differing from solids and liquids only in the amount of heat they contained. This fundamental contribution—like his solution of the weight of calxes problem and his (incorrect, as it turned out) theory of acids—was widely accepted by defenders of phlogiston. On Newtonian principles, the particles of matter should compact themselves under the influence of gravity. But for fluids and more strikingly for gases, something kept that from happening. The well-known tendency of substances to expand when

heated and grosser experience with melting and boiling made heat in some form the obvious candidate to account for changes of state.

As mentioned in note 5, Lavoisier (like Black, Franklin, and other nearly everyone else save Rumford and perhaps Cavendish) favored the notion of heat as a tenuous, pervasive fluid into which other substances were dissolved. But Lavoisier repeatedly allowed that some other specification of the physical basis of heat might be more correct. Solids normally contained little of this latent or fixed heat, fluids more (so that their particles were separated sufficiently to allow them to slide readily), and airs contained enough to wholly separate the particles of a substance, allowing it to expand to fill whatever space was available.

11. It is often claimed that critical use of the balance was a contribution of Lavoisier. Lavoisier indeed exploited the balance to great effect, but so did Cavendish and Scheele. Even Priestley, who was notorious for the number of qualitative errors he managed to make, was generally impressively accurate in quantitative work. A century earlier Boyle (in *The Sceptical Chymist*) remarks, "Chymists do universally take it for granted that in distillations carefully made, the matter that passes into the receiver . . . together with the remains, or *caput mortem,* amount to just the weight that the entire body had before distillation" (quoted in Partington 1961, 500). With the rise of pneumatic chemistry it was inevitable that this widely held intuition would be extended beyond the closed systems involved in distillations: not merely in the sense that it became inevitable that someone would discover the usefulness of meticulous use of the balance, but in the stronger sense that this development is hardly a distinct thing from the rise of pneumatic chemistry.

12. Musgrave and others have argued that Lavoisier did not really offer a simpler theory, since he needed to postulate that oxygen gas was a compound of elementary oxygen and matter of heat. So a functional equivalent of phlogiston was just transferred from the combustible to the oxygen, yielding no improvement in economy. But references to some functional equivalent of latent heat (under such labels as "matter of heat," "latent heat," or Lavoisier's eventual label, "caloric") are routine in the phlogistic accounts, since defenders of phlogiston as much as Lavoisier needed some such notion to account for the gaseous state. Cavendish, Priestley, and others speak of inflammable air as a compound of phlogiston, heat, and water. No one contemporary with Lavoisier suggests that his theory introduced a new substance in place of the vanished phlogiston. And of course all accepted oxygen (under some label or other). From our own point of view, the claim has a point. Logically, it was not clear—today we would say it is false—that Lavoisier was right to suppose that all the heat came from the oxygen gas. But it is wholly anachronistic to raise this as an issue relevant to positions taken in Lavoisier's time.

13. Cavendish was the one figure in the story whose talents seem entirely on a par with Lavoisier's. As the original discoverer of the composition of water, he had good evidence in hand for two years before Lavoisier heard of it, but was not moved to any doubt about phlogiston. For Cavendish we have a ready explanation. Since he was also the discoverer that phlogiston could be isolated (as inflammable air), he could hardly avoid some tendency to look criti-

cally at an argument that implied his discovery was only a blunder. But Lavoisier's bias would go the other way, with consequences that will be the focus of chapter 5. The fullest account, by far, is Wilson (1852, [1975]).

Chapter 5

1. Cavendish also very carefully and skillfully resolved the puzzle of why, under some conditions, the product of inflammable air + dephlogisticated air was a weak acid. He showed that a trace of acid came from the inevitable contamination of oxygen samples by the nitrogen that makes up 80% of the atmosphere. Cavendish's account is remarkably complete and accurate, and it provides one of a number of instances in which Priestley's lonely continued opposition to phlogiston in the 1790s was based on refusal to accept not only some of Lavoisier's results but also those of onetime allies such as Cavendish and Kirwan.

2. Carbon dioxide to us, but "aerial acid" at the time. When dissolved in water, it yields carbonic acid, which meant that in the presence of moisture, tests with litmus and so on would indicate an acid.

3. Priestley could miss the significance of the dew that formed in his combustion chambers, since for someone bound by phlogistic intuitions, the notion of water as a common component of gases was congenial. For example, varying water content might account for the varying weight and other properties of inflammable air from metals versus heavy inflammable air from other sources [4.6]. It could also account for the appearance of dew that Priestley and a number of other chemists had noticed after some combustion experiments. The agreeable character of this notion showed itself in the considerable effort Priestley made to prove that water could be transmuted into air. Priestley at one point had prepared a report for the Royal Society on experiments purporting to demonstrate that water indeed could be converted into air. This was done using earthenware jars supplied by Josiah Wedgewood. But when Priestley wrote to Wedgewood about his success, Wedgewood warned him that earthenware jars are slightly porous, so that he was probably expelling steam (under pressure) through the pores, after which atmospheric pressure would force air back into the jar, relieving the partial vacuum.

After 1783, the conjectured presence of water in airs became a central point for phlogistonists, as will be discussed shortly.

4. An important further element of the story (elided here) concerns the role of acids. The defenders allowed that burning phosphorus and sulfur, for example, involves absorption of oxygen, and yields their acids. Nevertheless, they continued to usually speak as if acids were elementary, so that sulfuric acid + phlogiston → sulfur, nitric acid + phlogiston → nitrogen. On this scheme fixed air (carbon dioxide) was elementary, and required the addition of phlogiston to become charcoal.

5. Priestley to Josiah Wedgewood, September 16, 1782.

6. The paper appeared in the Academy's *Memoirs* for 1783. But the volume appeared only after the usual delay, in 1786. The fall 1783 paper appeared in the *Memoirs* for 1781, published in 1784. A fairly common confusion

in the secondary literature is among the dates at which Lavoisier wrote a paper, read it to the Academy, actually published it, and the nominal date of the publication. Neglecting these distinctions considerably blurs the lines Perrin (1987) draws in his classification and the similar lines stressed here.

7. See the tabulation in Perrin 1988, table 3.

8. Kirwan 1789, 7. Kirwan's defense of phlogiston was dated 1784. Lavoisier and his collaborators published it in a French translation, with detailed replies. The result was then republished in an English version, with responses from Kirwan. All this provides our best evidence of how the issue was seen and argued at the time when the issue was still in dispute.

9. Lavoisier's point here is that Kirwan (and in their various ways Priestley and other defenders) supposes that the phlogiston once claimed to escape now remains with the calx, but bound to the oxygen absorbed during calcination [5.3]. So how could flame be escaping phlogiston? The defenders might "quine" this difficulty (hence still invoke phlogiston to account for emerging flames when, for example, iron burns in pure oxygen) by imposing the further assumption that the calx is able to draw the resulting water or fixed air (taken to be the product of the visibly escaping phlogiston) back into itself. So, as usual, the difficulty is not essentially a logical one. It is always possible to quine. But as the complications mount, sustaining the appeal of the original simple intuition that flames are some escaping substance depends more and more on an entrenched habit of mind.

Lavoisier also writes: "This supposition [that metals et al. contain inflammable air] is by no means proved; and I will say even more, that it is not at all probable: but, even if it should be . . . demonstrated . . . this would not at all invalidate the proof that the air [meaning oxygen] furnishes the caloric [heat] and light during the act of combustion" (Lavoisier's comment, in Kirwan 1789, 22). Lavoisier's confidence here presumably comes from the "steel wool" experiment, where flame is visibly rising from the metal while oxygen is being absorbed. A puzzle for the Lavoisier argument is to explain where the heat and flame comes from when gunpowder is exploded. I have not come across an explicit discussion by Lavoisier, but presumably he would attribute the effect to some exceptional capacity of gunpowder to store heat in a latent form. The existence of latent heat somehow stored in solids was recognized by then, and seen as a way to explain the large variations in the capacity of different materials to absorb heat for a given rise in temperature.

10. Each of Lavoisier's major adversaries had his own variant of the phlogistic case—a point that of itself indicates the difficulty of framing an account that seemed plausible to anyone but its sponsor. I try to give a reasonable though certainly incomplete sense of the range of positions as the debate proceeded during the 1780s. Discussion beyond the 1780s is not only unimportant in terms of changing anyone's mind but inconsequential also in the sense that the mainstream community was no longer interested, nor were there any interesting new arguments. Priestley, the sole holdout of any consequence, had by then moved to America. He increasingly relied on experimental results that no one else reported.

11. If inflammable air from charcoal and other organic materials was a

variant on inflammable air from metals (both were taken to be primarily phlogiston), one could explain various results by the apparent ability of phlogiston + oxygen to form either fixed air (carbon dioxide) or water, depending on conditions. This claim, and the counterclaims from Lavoisier, played a considerable role in the argument of the defenders.

12. Since Lavoisier provided a clearly plausible alternative explanation that denied the existence of phlogiston, the disengagement of inflammable air from metal dissolved in acids obviously no longer provided a prima facie case that phlogiston existed. The bare fact was as compatible with the denial of phlogiston as with its assertion.

There are large presumptions in any claims about simplicity, plausibility, and the like as something independent of the subjective response of particular individuals. But in terms of everyday discourse (in contrast to philosophers' talk) there is nothing that should be controversial here, as indicated by the fact that leading defenders of phlogiston themselves perceived Lavoisier's theory as "almost as simple" (Cavendish) or even as simpler (Kirwan) than their own. If even defenders fully committed to phlogiston saw things that way, it is scarcely surprising that someone *not* firmly committed to phlogiston would be even more emphatic on that point.

As I have mentioned in chapter 4, note 5, a common counterargument here (that Lavoisier replaced phlogiston by caloric, and so was not more economical) is wrong, because phlogistonists accepted some functional equivalent of caloric (heat), in addition to phlogiston. The claim that Lavoisier merely replaced phlogiston with caloric seems to be entirely a recent invention. So far as I know, no one in the eighteenth century said that, or anything that could be reasonably construed as implying that.

13. Scheele died in 1784, so we cannot know whether he would have converted to the new ideas soon after 1783 or, like Priestley, would have held on for years. Cavendish probably did go over to Lavoisier's position, but no positive record of that seems to have survived. Certainly he did not continue to actively defend phlogiston, and there is a bit of indirect evidence that suggests he did explicitly abandon it. In their replies to Kirwan, one of Lavoisier's collaborators notes that Cavendish was the originator of the claim that increase in weight of calxes is due to formation of water attached to the calx. He then says that, although Kirwan was continuing to defend that sort of view, Cavendish himself abandoned it. If so, apparently Cavendish had abandoned phlogiston sometime between 1784 and 1788. Neither Kirwan nor the English editor of the exchange challenged the French claim.

14. Here, as is unavoidable, the debate is much simplified, so that the case seems starker than it would to a reader *today* of (say) the exchanges between Kirwan and Lavoisier. But a serious participant at the time the issue was really in contest would have become very familiar with the arguments on all sides, and the advantage to Lavoisier would reappear: from the massive defections Lavoisier won starting in 1784 we can infer that such a striking superiority did appear. The conversion lagged by perhaps a couple of years in England, where local sentiment naturally favored Priestley and Cavendish.

15. The reference here, of course, is to the major epicycles that govern the retrogressions of Ptolemaic planets and determine the spatial structure of the Ptolemaic world. They disappear in the Copernican scheme. A full discussion of this issue comes in chapter 7.

16. The original proposal to the National Science Foundation for this research promised an analysis of phlogiston, on the grounds that it seemed especially likely to be fruitful.

Chapter 6

1. Among the familiar proposals for dealing with this puzzle is the claim that gamblers used an asymmetric form of dice (astragali) and so did not have a ready model for equally likely cases. But dice essentially like those used today date back at least to Babylonian times. Another familiar explanation is that randomizers were used in religious divination, so that it would be indecent to subject chance to cold analysis. But recreational gambling has always been common, and ancient comments on it that have come down show no hint of religious awe inhibiting discussion.

2. Fermat-Pascal correspondence, which is reprinted in appendix 4 of Florence David's 1962 book. References to this correspondence are noted as FP, using the paging in David 1962.

3. Hacking's 1975 book has been much admired, but his claim that the modern ordinary language sense of probability appeared only when a calculus of probability did has elicited numerous counterexamples. See especially Garber and Zabell (1979).

4. David (1962) provides a discussion of this and most other historical points mentioned here, and provides a translation of Galileo's paper.

5. Cardano's guide—dating from the 1540s—is not a convincing counterexample. There are two oddities that raise doubts about just what Cardano actually wrote. First, in contrast to Pascal, Fermat, and Huygens, who are all clearly aware they are doing something new and interesting as mathematics, there is no hint of any of that in Cardano. His analysis comes as a few pages in the middle of a booklet of advice to novice gamblers (roll the dice vigorously, he says, and luck will be with you), which he never bothered to publish. Cardano, who was ordinarily ardent in asserting claims to priority, here is indifferent to a remarkable achievement. In his autobiography, he mentions in passing writing something on gambling, but nothing that indicates he considered it to contain anything original or important.

A second oddity is that the text repeatedly gives a simple but incorrect analysis, followed by a correct analysis. Ore attributes this (1953, 164) to Cardano's "method of composition," where "the first erroneous attempts are included as well as the final correct arguments." But this is certainly a very strange method of composition: in general a writer finds it almost physically impossible to correct an error without crossing out the mistake it corrects.

Finally, there is the curious fact that the text we have was only published (1662) after its most striking ideas were available elsewhere, and in fact it was published in France (Lyons), where an editor of Cardano's mathematical papers would be especially likely to be aware of what Pascal and Fermat had done. A reasonable conjecture of what is going on here is that someone circa 1660 wrote out the correct analysis as a commentary on Cardano, and that somehow this material was included in Cardano's printed text.

It should also be noted that the text as published in 1662 deals only with odds (not explicitly probabilities). Ore (1953) reframes the arguments in modern terms. Gould's translation (printed as an appendix to Ore) also changes odds to probability, but only for the case of equal odds, which is translated as $p = .5$. So the usage in the text is just that of Fermat and Pascal, not reaching the more nearly explicit use of probability as quantitative found in Huygens's little treatise (1657).

6. See, for example, Sagredo's usage (Galileo 1638, 155), where he casually says that a falling object attains a degree of speed with each unit of time (pulse beat). If he means this quantitatively, he is casually assuming the very thing that Salviati (speaking for Galileo) presents a few pages later as a highly counterintuitive discovery, and which he has Salviati argue carefully to *persuade* Sagredo of its plausibility.

7. Drake (in his translation of Galileo 1638, xxi–xxv) is emphatic that Galileo would not have used a single number to measure speeds, since no correct or well-defined ratio could be formed between two different kinds of magnitude (distances and times). Hence (Drake argues) Galileo always speaks in terms of proportions, such as $d1 : d2 :: s1 : s2$. Indeed, "degrees of speed" is sometimes spoken of in a way that seems merely qualitative (see note 6), with quantitative arguments restricted to the use of proportions. In other places, degrees of speed seems to be quantitative in just the way we would speak of one speed being twice as great as another (for example, 159ff.). Perhaps the explanation is that Galileo himself has reached what I call in the text inverse counting (so he finds it comfortable to think of speed as a quantity), but to make the argument easier for readers, and perhaps also because he does not yet feel wholly confident this is okay, his detailed argument is always in terms of Euclidean proportions. A few decades later, what appears to be a parallel situation arises, as Newton avoids explicit use of the calculus in the *Principia,* presenting everything in terms that at least look like classical geometry. See also the comments in note 6.

8. Heilbron (1990) wittily reflects the process in the course of a book review: "Fischer takes pains and, it appears, pleasure, in deflating Napoleon's calculation of his mathematical abilities. Fischer himself deserves credit for a unique contribution to mathematics, in the form of the quantification of footnotes. His begin at -44, ascend to zero at the last annotation of his introduction, and finish at 1295. The negative footnote has made its appearance to complement the irrational and imaginary ones with which we have long been familiar."

9. Why not (a reviewer has asked) say that this establishes a habit of mind, rather than tames an anomaly? If we use "habit" to mean an entrenched response, looking for how a comparative quality might be made quantitative is the sort of thing I have in mind for the label *habit.* Seeing money as familiar,

concrete units to give prices is an example of a recognizable pattern in the repertoire, not of what I have gotten into the habit of calling a habit. Perhaps the point can be made sharper by recalling the discussion in chapter 1 of the meaning of a word in a particular context: familiarity with a particular meaning of a word involves (in the way I have been talking) having a pattern in the repertoire. But using a particular meaning of the word in certain contexts (of several meanings of that word that are in the repertoire) is a habit.

10. "Mathematics is for mathematicians," Copernicus says in the introduction to *De Revolutionibus,* in the course of warning theologians not to make fools of themselves with respect to a matter they are not competent to understand.

11. Geertz (1973) gives a vivid description of setting of odds in side bets on cockfights in Bali, which takes the form of a two-sided auction.

12. It might be supposed that fair price can only follow the more general notion of price. That is plausibly but not necessarily so. Price might be fixed by custom at first, with any departure from that seen as a kind of theft. Then price would be fair price. Only later might the idea emerge that price is something that can properly vary from transaction to transaction, hence that might be bargained about.

13. I want to mention again that the account here conflicts not only with Ore's charitable reading of Cardano but even with a less generous one, since on the argument as I have developed it here the critical step is (1) the move from odds as a price that emerges from bargaining to odds as something that can be calculated, not (2) the step from odds (or prices for gambles) that can be calculated to seeing probability as a number between zero and one. See the discussion in section 6.5. Several variant versions of the emergence of probability could be framed consistent with the main barrier argument, sharing a family resemblance, since all turn on the absence of habits of mind that can prompt inverse counting. I give what seems to me the most plausible version. If you are not troubled by the suspiciously late publication date of Cardano's analysis (note 5), then the story would be tied to the novel extensions of *number* a century earlier, with which Cardano was much involved. At the other pole, where it is the appearance of probability explicitly as a number between zero and one that is crucial, the emergence comes with Huygens (as David argues) or even not until Bernoulli and others were making their contribution at the turn of the next century.

14. To give the simplest case: with two dice there is one combination that yields a total of two (1,1), and one that yields a total of three (2,1). But threes are twice as common as twos. If the dice are unmatched (say white and red), the explanation involves only direct counting: we get a two only from white 1, red 1. But we get a three from both white 1, red 2 and red 1, white 2.

15. FP 231. In postemergence language: $(1 - p)^4 = (5/6)^4 = .482$ is the chance of failing to make a particular point in four throws.

16. The numbers here are still small enough that it would not be an overwhelmingly onerous task to list all the possible sequences of outcomes, and count the fraction that are favorable. Or perhaps de Mere knew some elementary

combinatorics, which does not involve inverse counting and which dates back long before the emergence of probability. Then he could quickly calculate the total number of possible sequences of four throws as $6 \times 6 \times 6 \times 6 = 1296$, and the number of sequences excluding 6 as $5 \times 5 \times 5 \times 5 = 625$. So the favorable cases are $1296 - 625 = 671$, and the fraction that include a 6 is then 671/1296, or almost .52. Notice that this does not require that the person doing the calculation sees a probability: the calculation is effectively $(5/6)^4$, but cognitively $5 \times 5 \times 5 \times 5 / 6 \times 6 \times 6 \times 6$, which involves no probabilities.

17. $\tfrac{35}{36}^{24} = .509$, $\tfrac{35}{36}^{25} = .495$. It takes 25 throws to make betting you will get a twelve advantageous, so de Mere was correct in his confidence of where the advantage lay.

18. Without some elaboration, the problem has no unique solution. The most common actual rule is that an incomplete game is no game, so that each player would get back his original stake no matter what the score. But in context, we can see that the problem was understood as one of fair division, so that neither side gains an advantage from the interruption, and with any particular point equally likely to be won by either player. That there are versions that talk of the price to be paid by a bystander who wants to take over someone's position suggest attempts to make that interpretation less open to quibbles.

19. Pascal and Fermat were not correcting a generally accepted answer. Awareness that $3 : 1$ is wrong could have arisen, for example, by discussing an extreme case: say where ten points are required, and I have nine while you still have none. Just counting possible decisive sequences, there are ten that favor me, that is, where a head shows up on any of the next ten tosses. And there is just one where you win: where tails comes up ten times in a row. So the fair division, on this count, would be $10 : 1$. But now a conflict with common experience is sufficiently striking to tell any gambler that the chance that tails will come up ten times in a row is much less than one chance in eleven (in fact it is one in 1024). Gamblers remember extraordinary runs of luck and have some sense of how often they occur, as every poker player is aware than in an evening's play it is very unlikely that a straight flush will be seen.

20. David (1962, 90–94) even suggests that Pascal himself did not have things quite right, and was covering his own confusion by references to Roberval. But this seems to be a misreading of Pascal's comment.

21. De Mere and Roberval could calculate where the advantage lay in betting whether a given point would appear in four rolls of a die, which logically is just a slightly *more* complicated problem than calculating where the advantage lay in a points-with-players problem where one player needs one win and the other player four. To solve the dice problem de Mere has to calculate the fraction of winning cases, and see whether it is more or less than half, where the chance of getting the point on any one throw is one in six. The player who needs one win in the points-with-players case has one chance in two on each opportunity (supposing, as indeed was supposed, that neither player had an advantage). So the "players" calculation is a little simpler.

22. See the counterargument, note 5 (on Cardano).

23. Suppose one side needs two wins, as against three by the other. Then on the play before the issue is decided, someone needs one win, and the other one or more. But (by now) we know how to calculate the value for one-win cases. The player who needs two wins needs one win to get to the now-solved one-win case. So Pascal can work out the value for the two-win case. And so on.

24. It is worth emphasizing that these cases are different. Cows are naturally countable. Dollars are a construction for making value quantitative, and hence are an outstanding example of inverse counting, but one so familiar from so early an age, and in the form of tokens that are naturally countable, that its artificial character is long forgotten.

Chapter 7

1. There is an unavoidable ambiguity in talking of trailing or leading the Sun here. Relative to the annual cycle of the Sun through the zodiac, when the planet is leading the Sun it is seen as trailing the Sun during the daily rotation (so we see the planet rising after the Sun has set). For as already mentioned in the text, the annual motion through the zodiac goes west-to-east. The Earth also spins on its axis west-to-east. But we perceive this as a daily motion of the heavens east-to-west.

2. On both Copernican and Ptolemaic models, Venus (like Mars) should vary in distance from the Earth by a factor of about six over the course of its orbit. But in contrast to Mars, there is no apparent change in the brightness of Venus as a function of its computed distance. In his preface to Copernicus, Osiander cites the Venus puzzle as showing why no astronomical system can be taken to be true. (Brightness does vary with other considerations, but not with position relative to the Sun within the limits of naked-eye observation.) Galileo was able to provide a simple Copernican explanation, since his telescope showed that Venus moves through a complete cycle of phases (like the Moon). Hence when it is near the Earth, very little of the lighted surface would be seen, and when it was full it would be at its greatest distance. The joint effect accounted for the absence of noticeable variation in brightness with distance. But Galileo guessed this only a few weeks before he was able to make the observation (as discussed in *Patterns* [12.11]). Until then no one, the Copernicans included, seems to have noticed this possibility. If the issue had been raised, a possible Ptolemaic response would have been that the planets are translucent, since Venus appears spherical to the naked eye even when the telescope shows it as a thin sliver. It would not have been an unreasonable conjecture (given what was known at the time) that the planet is lit up by the Sun shining through it more brightly when close to the Sun, hence offsetting its greater distance from the Earth.

Venus as seen through the telescope would exhibit identical phases on either Copernican or Tychonic assumptions. So Tychonic as well as Copernican astronomers would have been motivated to make something of this explanation had it been conjectured. On the other hand, if (as the evidence reviewed in Margolis 1991 suggests) Ptolemaic belief had already been severely eroded by the end of the sixteenth century, the Venus puzzle would not have been relevant to what was actually under active dispute (that is, Copernicus versus Tychonic astronomy, not either versus Ptolemy).

3. On refutation of Duhem's (1985) claim that astronomers were only saving appearances, see Ragep 1990, Evans 1991. For direct contrary indications, see especially Ptolemy's remarks on the relation between mathematical astronomy and natural philosophy in the opening passages of the *Almagest* and his explicit discussion of the physical structure of the universe in the *Planetary Hypothesis* (Goldstein 1967).

4. The primary orbit shown in diagrams like figure 7.2 must be understood as being a mathematical locus, not a physical thing. The epicycle is carried by a disk (like the old 45-rpm records) or sphere (the three-dimensional analogue of that), which rotates in place relative to its own center. It can be carried around in a circular motion by a larger entity, but it does not move from place to place on its own. So the planet is carried around by the rotating epicycle, and the epicycle is carried *inside* the rotating physical deferent (the disk). Figure 7.6 shows the physical setup for a Ptolemaic model, with the rotating deferent carrying the epicycle inside it, while the rotating epicycle carries the planet. From physical experience with persisting motion, it is not surprising that rotation in place (like a potter's wheel), not translational motion, would seem peculiarly appropriate for eternal motions in the heavens.

The epicycle in diagrams like that of figure 7.6 is conventionally shown as a disk. But notice that only the rim of the epicycle, which carries the planet, plays a role in the cosmic machinery. This is an important point, since I will eventually comment on the possibility of a "solid spheres" version of the Tychonic system, where it will be essential to take note of the possibility that the epicycle could just as well be a ring instead of the disk that is ordinarily shown. We will eventually (chapter 8) have occasion to consider *inverted* models, where the larger orbit rides on the smaller. Then the epicycles would surround the center of the deferent, rather than sit within the deferent. This violates familiar conceptions of a solid-spheres model, but is perfectly consistent with the logic of such models.

5. This means closest, given some location of the center of the epicycle relative to the Earth, not necessarily in absolute distance, since (as will be discussed later) another source of variation in Earth-Sun distance is associated with the location of the planet in the zodiac.

6. Here is where a reader who wants to be confident about these points should start to draw figures. Use a compass to try a figure with a relatively large ratio of epicycle to deferent radius, and another with a smaller ratio. Suppose that the periods are set: say the epicycle rotates at just half the period of the deferent, as is a good approximation for Mars. Once you draw the figures, or think through the situation with the help of mental imagery, it is easy to see that the bigger the epicycle is relative to the deferent, the more marked the variation of motion as seen from the Earth.

7. All references to direction refer to conventional reading of diagrams. Obviously the Sun is never due north of the Earth in the physical world.

8. After Copernicus, this coincidence had an explanation. Since the epicycles are seen to be only a reflection of the Earth's own motion relative to the Sun, of course these appearances coincide, as the motions of multiple reflections in a pair of mirrors coincide. The startling simplicity of this explanation is emphasized by all the early Copernicans.

9. Wouldn't it be simpler still for Ptolemy and his predecessors to conjecture the Tychonic or Copernican account? But the Tychonic scheme not only does involve epicycles, as will be emphasized in the next chapter, but unless you are in a position to apprehend the Tychonic setup from familiarity with the Copernican setup, the way Tycho accounts for retrogressions is not easy to see. An early astronomer would not easily see the Copernican possibility from the apparent variation in distance of a planet (becoming brighter at retrogression) for the same reason that, if you are on a train that starts moving gently from the station, it will at first seem that the train next to you is moving to your rear, not that you are moving forward. Unless we are jolted into consciousness of our own motion, our perception is that we are standing still and the external thing is moving. It requires some extraordinary wrench from usual perceptual habits to prompt an astronomer to see the possibility that the Earth is coming closer to a planet at retrogression rather than the planet coming closer to the Earth.

10. In a Ptolemaic world, Mercury is sometimes just beyond the distance of the Moon, whose parallax can be measured. But we can see the Moon (when full) all night long. So we have a baseline of some twelve hours for observation of parallax. Mercury can only be seen for a short while at sunset or sunrise, and when it is closest to the Earth, it is too close to the Sun to be seen at all.

11. There was an easily available notion of average sidereal (that is, with respect to the fixed stars) period, obtained by counting the number of years between returns to the same position in the sky at the same location on the epicycle. The number of retrogressions over that interval then gives the mean time between retrogressions. But the variations in interval, size, and shape remain unexplained.

12. That the epicycle idea readily prompts disbelief is apparent on the record. A habits-of-mind conjecture about what might account for that turns on the point that experience in the world is that an object carried round is pressed against the periphery (by centrifugal force). And there would be an intuitive presumption that if nothing can be seen, probably there is nothing to be seen. And there would be an intuitive aversion to introducing new entities as odd as huge, invisible wheels in the sky.

These intuitions are not invincible. But someone who has not wrestled with the problem of making a model of the planetary motions come out right without using epicycles easily supposes that, if the astronomers were sufficiently clever, they should be able to account for the motions without epicycles—or at least that if human astronomers are incapable of that, surely God could so manage. Only thoroughly expert astronomers would be fully aware of the many precise ways in which observations neatly fit expectations of what would be seen if the epicycles were real, which is what generates a sense that such objects must indeed be real.

13. Hence Ptolemy called this effect the anomaly with respect to the zodiac, while he called retrogressions the anomaly with respect to the Sun.

14. Wilson (1975) argues that Copernicus may have been led to his heliocentric view because the additional epicyclets made him want to simplify his models in some other respect. But the Arabs made this move with no hint that they saw the epicyclets as anything but a solution to an anomaly in Ptolemy. Nor is it psychologically very plausible that this secondary concern would explain, after fourteen centuries, why an astronomer finally came to see the heliocentric possibility. In any case, if Copernicus was so moved, we presumably would find exploitation of this justification in his defense of what he conceded would strike readers as an "absurd" claim that the Earth was a planet. And in fact there is a passing allusion to the point in *De Revolutionibus* (introduction) and a very brief but explicit mention of the claim in Rheticus's preliminary narrative. But claims that this was important for Copernicus come mainly from the prominent position the role of the equant is given in his first sketch of his theory, dating from about 1610 (*Patterns* [12.8]). This could reflect both what Copernicus thought would catch the attention of the astronomers in Krakow whom he wanted to engage, and also his subjective sense that the equant problem indeed was important for his discovery—as indeed it was, in a serendipitous way, on the account here. It could also explain why the equant point gets a more emphatic remark from Rheticus, who might have been shown the early sketch during his long visit with Copernicus, than it does from Copernicus himself when he finally let his book go to press. There Copernicus, parallel to the contrast between prominence in the *Commentariolus* and a bare passing mention in the cosmological discussion of *De Revolutionibus,* drops any suggestion of a logical link between the two ideas.

15. On the record no Ptolemaic astronomer made this linkage explicit, and it would have taken a difficult move to do so: in the language of this study, there was a barrier against doing that, as will be discussed in chapter 8. But the linkage motion is tacitly built into the Ptolemaic models.

16. The usual manner of speaking—and presumably the usual belief among premodern astronomers—refers to spheres. But as Ptolemy himself mentioned (Goldstein 1967), all that is required for deferents are disks. Since disks are easier to visualize and to capture in diagrams (such as figure 7.5), that is what a reader will do best to think about. But following customary usage, I will usually refer to spheres. Parallel to this, the epicycle need only be a ring, not a full disk, a point I will have occasion to use later.

17. Recall that physically the deferent is a sphere or disk that carries the epicycle, as shown in figure 7.6. The deferent shown in diagrams like figure 7.2 or 7.3 is not a physical object, but the mathematical locus of the center of the epicycle as it is carried around by the physical deferent. References to R (radius of the deferent) are to that mathematical locus. If you spend a moment or two thinking of the relation between the way things are shown in figure 7.2 or 7.3 and the way the same situation is shown in figure 7.6, you will see that

nothing very difficult is involved in this. If you don't do that, but just look at the words without attending to the diagrams, then the situation is likely to seem terribly difficult. So attend to the diagrams.

18. Ptolemy does not give a further argument for the ordering that was available to him but apparently not noticed. If the Ferris-wheel motion is deducted from the epicycles of Mercury and Venus, we get the periods of 88 and 225 days familiar to us from Copernican astronomy. An ordering governed by the speed of the p-orbits then gives just the ordering Ptolemy wants to show is reasonable. But Ptolemy and all succeeding Ptolemaic astronomers neglected this argument. So on the face of things, they did not notice, or at least did not see as physically meaningful, the linkage way of articulating how the models work. Rather, for Ptolemaic astronomers the p-orbits of Mercury and Venus were just the observed periods of 145 and 584 days. The issue plays a role in chapter 8.

19. This scheme is laid out by Ptolemy and almost certainly must originate with him. There was no hint of varying distances to the planets in Aristarchus's suggestion that the Earth might orbit the Sun (*Patterns* [11.6]), which is not surprising since this preceded the epicycle idea, though not by very long. At least until the epicycle idea arrived, the starry vault was easily and probably usually seen as essentially flat, with the planets as well as the fixed stars set in the far background behind the Sun and Moon. Occasional numerological speculation about the distances to the planets would not be sufficient to challenge that, though such speculation does show that the idea that the planets vary substantially in distance is not counterintuitive. Experience favors seeing lights that are of different brightness as being at different distances. For the fixed stars that is offset by the converse intuition with respect to lights that move exactly together. But the planets move independently. On the other hand, before *effective* abstract models (like figure 7.2) were available—that is, before Ptolemy—intrinsically spatial models (like figure 7.6) could not emerge. And until such models took hold, nothing even in the specialized experience of astronomers would lead to the entrenched nested-spheres habits of mind that took hold once the scheme being sketched here took hold.

20. The parallax of the Sun could not be directly measured. Not only is it far smaller than for the Moon (of course), but for the Moon an astronomer could watch for a shift relative to nearby stars, and do that fairly accurately even with naked-eye observation and despite a need for an adjustment to take account of the motion of the Moon over the course of a night. But the Sun hides nearby stars in its glare, except for the rare and momentary opportunity at times of total eclipse. Indirectly (using observations of eclipses) parallax for the Sun can be estimated. See Van Helden 1985 for details. The results are radically unreliable (angles must be measured much more precisely than available instruments could justify). That allowed wishful thinking to shape the observations, rather than observations to constrain wishful thinking. But it led to such a pretty story (sketched in the text) that the Ptolemaic estimate was never questioned. Even Tycho, with his commitment to reestimating all the ancient parameters with his enlarged instruments, never questioned Ptolemy's number here, nor did

Copernicus. Ptolemy in fact misestimated the distance by a factor of twenty. The whole argument is absurd in terms of what we know now and even in terms of what they logically could have known. But it was not at all seen so until after Ptolemy had been abandoned on other grounds.

21. See Kuhn 1977 on what constitutes "reasonable agreement" with theory.

22. Kepler's Rudolphine tables—the result of his abandonment of the whole Ptolemaic apparatus, which had been carried over by Copernicus—was an improvement. But by the time these tables were available (1630), there were no more Ptolemaic astronomers. Kepler's work was possible only on the basis of the order-of-magnitude improvement in naked-eye observations provided by Tycho's data. But it is naive to suppose that if only there had been a Tycho centuries earlier, there could plausibly have also been a Kepler to see that elliptical orbits could yield more exact models than the Ptolemaic circles. Kepler had enormous advantages over any astronomer who was not a Copernican. Since the major epicycles had disappeared, doubts could come easily about the reality of secondary epicyclets. That helps account for Kepler's early abandonment of the epicyclets that Copernicus used to give a physical account of Ptolemy's equant. And again, very early in his career, before his connection with Tycho, Kepler realized that the epicycles that Copernicus as well as Ptolemy used to account for variations in latitude also were unnecessary. (This leads to the one critical comment Kepler made on Copernicus, that on this point he paid too much attention to Ptolemy and not enough to nature.) Even so, the path to the ellipse was difficult for Kepler, and tenuous as well. His ellipses became unambiguously convincing only after Newton provided deeper foundations for the idea.

23. See *Patterns,* chapter 14, and further detail in Margolis 1991.

24. That is, until now you could think of R *either* as the radius of the deferent orbit *or* as the radius of the larger orbit, since the deferent orbit in the standard Ptolemaic setup is always the larger orbit. But if we want to allow for inversion, we need to settle things one way or the other. The convention we are choosing is that R is the radius of the larger orbit, whether or not it is the deferent, and r is the radius of the smaller orbit, whether or not it is the epicycle.

25. As an exercise, figure out why it is impossible in a world of Ptolemaic nested-spheres that we would find a planet for which the radius of the s-orbit is the same as the radius of the p-orbit. What would that situation imply in a Copernican world?

26. Even before Copernicus, it seems to have been congenial for someone who was *not* a trained astronomer to see these planets as heliocentric. But not so for someone who was. The point, which is important for understanding how recognition of the Copernican possibility could have been delayed for so many centuries, is discussed in detail in *Patterns* (cahpter 12) and will be reviewed here in chapter 10.

27. That Ptolemaic belief began to be displaced by Tychonic or Copernican belief by the late 1580s is not a controversial claim, but this shift in belief is commonly explained as driven by Tycho's observations of the nova ("new

star") of 1572 and of the great comet of 1577. I comment in some detail in *Patterns* [13.7] on why that is not an adequate account of this radical shift in astronomers' belief that had never been seriously threatened in fourteen centuries. A little more detail will be added in chapter 10.

Chapter 8

1. On Kepler's (and Galileo's) dismissive attitude to the Ptolemaic system even before the discovery of the phases of Venus, see Margolis 1991.

2. Various further arguments are commonly cited to account for the decline of Ptolemaic belief before the telescope. Tycho himself argued that the paths of comets showed that the Ptolemaic nested spheres must not exist. The comets, this argument amounts to saying, appear to move as though the nested spheres are not there; hence they are not there. A second argument points to the unambiguous ordering of the planets if they are heliocentric, as holds for both Copernicus and Tycho.

But the comet argument seems to have been taken to be strong in the secondary literature mainly because we know of no one who answered it. Yet common experience makes clear that merely because an argument is a good one will not preempt disbelief and efforts to answer it if there are people around motivated toward disbelief. And in this case, it is easy to think up perfectly good counterarguments (*Patterns* [13.7]). More fundamental is that, even if we accept Tycho's argument, it is not a positive argument for preferring his system to Ptolemy's. In Tycho's system, the planetary orbits are epicycles centered on the Sun. Any explanation Tycho might give for how the Sun carries the epicycles of the planets around in his system either works conspicuously better for the Copernican system, or could be adapted to explain how the Ptolemaic epicycles are carried around.

The Copernican (and Tychonic) unique ordering of the planets also is hard to credit as a *cause* of loss of belief in Ptolemy. Although an unambiguous ordering is implied by the Copernican or Tychonic systems, a (slightly different) ordering is also plausibly implied by Ptolemy [7.4]. A Ptolemaic astronomer would easily regard the stronger result that follows on the Copernican and Tychonic views as showing merely that the rival systems are rigidly wrong, in contrast to Ptolemy, who is plausibly right. Again, the absence of explicit counterarguments in the record is more plausibly a consequence of the absence of astronomers motivated to defend Ptolemy than a reflection of the irresistible character of Tycho's arguments.

Overall, we have strong grounds for confidence that Ptolemaic views had pretty well faded even before the advent of the telescope, but not because of the force of Tycho's arguments. The point of this chapter is to say how the basis for belief in Ptolemy had been eroded, quite apart from the claims for Tycho's system.

3. Lightman and Gingerich (1992) discuss this issue in a wider context, and this chapter is in part an outgrowth of discussions of the matter with Gingerich.

4. Perhaps yet another way of stating this will be useful. Recall (from section 7.2) that for each planet the ratio of the smaller to the larger orbit ($r:R$) can be fixed by observations. As long as one orbit for each planet can be fixed, the other is automatically determined. In the Ptolemaic scheme, this is worked out for each successive planet on the nested-spheres logic sketched in section 7.3. In either the Tychonic or Copernican scheme, the scaling is done all at once, by setting all s-orbits equal to the Earth-Sun distance.

5. See Galileo's polemic against Grassi, translated in Drake and O'Malley 1960. The point here is that Tychonic astronomy was always parasitic on Copernican astronomy, as discussed in some detail in Margolis 1991.

6. This is a touchy point, since so far as I know all modern commentators accept Tycho's claim. Nevertheless, what is required in principle of a proper solid-spheres model is (1) that everything fit together snugly (with no overlaps or gaps) and (2) that all motions be that of units rotating eternally in place. The planet has no motion of its own, but is carried round by the epicycle, the epicycle by the deferent [7.4]. You can think of the inverted models I am describing as like the old 45-rpm records, with a large hole in the center. The hole is large enough to accommodate the outer radius of the planet next closer to the center. The outer edge of the platter sets the size of the inner radius of the next planet. The p-orbit for each planet would then be a ring embedded in the platter, turning independently with its characteristic period. The locus of the center of the ring would lie inside the hole in the platter if $r:R < 0.5$, as in figure 7.9. It would lie within the platter (giving the apparent intersection familiar from Tychonic diagrams) if, as it is for Mars, $r:R > 0.5$, as in figure 8.1. But in neither case could there be any physical conflict, any more than you could bump into the equator. The intersection is only with a mathematical locus, not with a bumpable physical thing.

7. Comment on the Tychonic claims usually implies that his argument against the existence of solid spheres was both very strong and a direct blow to Ptolemaic belief. For a contrary view on the first point, see *Patterns* [13.6]. The second point is also mistaken; it implies that this argument provided Tycho with a cogent reason for dismissing Ptolemy, since Ptolemaic cosmology relied on the existence of solid spheres. The argument, however, would work only for a Copernican. Without solid spheres it was hard to see how Ptolemy's planets would carry out their complicated motions. But the same point holds for Tycho's planets. There exists no Tychonic explanation of how the planets would move *as if* their orbits were solid objects in some way rigidly linked to the Sun, though they aren't (since the heavens are not solid). But any explanation a defender of Tycho might give could be adapted to work for Ptolemy as well.

Tycho himself did not claim that his argument against the spheres damaged Ptolemy, only that, if there were no solid spheres, that removed the major objection against his alternative system. If there were no solid spheres, the intersection of the orbits of Mars and the Sun would not be a problem. As argued in note 6, this seems to be based on a misunderstanding of how the Ptolemaic physical models work. That Tycho would be capable of such a slip suggests that Copernican habits of mind (so that, like us, he no longer had ready intuitions about how things work in a Ptolemaic world) were already in place *before* Tycho proposed his argument. I will comment on that further as we proceed.

8. For a more detailed discussion of the simplicity of the Copernican system compared to the Tychonic, see Margolis 1991, section 2. Note that, in addition to eliminating the retrogressions (as physical movements), the Copernican scheme also eliminates the need for the linkage motion. On the other hand, many claims about the lack of a striking gain in simplicity from the Copernican insight simply confuse the analysis of retrogressions with the effects of what Ptolemy called the "anomaly with respect to the zodiac," compounding this with the mistaken supposition that the Copernican treatment of the equant was somehow a necessary part of his heliocentric argument. This is like an argument that a vacation in Tulsa is as expensive as a vacation in Paris, since the individual bought his wife a diamond ring while in Tulsa.

9. Discussed in *Patterns* [13.8] and later in this chapter.

10. *After* the invention of the telescope, when it became impossible to doubt that Venus was closer to the Earth than Mercury is, this Ptolemaic idea was visibly wrong. But as I have argued, there is plenty of evidence (see Margolis 1991) that the Tychonic system had already effectively displaced the Ptolemaic before that. Hence the evidence of the telescope cannot explain why the Tychonic arrangement so totally dominated the inverted Ptolemaic during the period when belief in Ptolemy was fading. Rather, the inverted possibility was not seen at all, and hence never was considered as an alternative to either Ptolemy or Tycho.

11. Why not "always" instead of "often" here? Because although logically not seeing x might be interpreted to be the same as seeing not-x (since on standard logic it is always the case that either x or not-x holds), cognitively that is not what happens. We often see neither x nor not-x, but something blurred, or uncertain, or we just don't notice anything in particular at all. Before the emergence of probability (chapter 6) people did not think probability was *not* quantitative; they just did not notice that it was.

12. Perhaps yet another reminder will be useful here that such processes are unconscious and pattern-governed, so that the active language here is not to be taken as a literal description of how choice is made.

13. Lightman and Gingerich (1992) provide a number of other examples, focusing especially on the striking case of the "flatness" issue in contemporary cosmology.

14. Since the only difference between the Tychonic system and the inverted Ptolemaic system concerns the scaling of the s-orbits, this point is not relevant with respect to the choice between the inverted system and the Tychonic. The problem is identical for either system. Neither has any relative advantage tied to this complication. In the Copernican system, of course, no such problem arises.

15. The relevant thought experiment here for a reader familiar with the Ptolemaic system requires forming a mental image of the planets in space. My own experience is that if I try to "see" the retrogressions in terms of diagrams

like those of figure 7.2 or an analogous Copernican diagram, then it seems about equally easy to do it either way. But if I try to envisage the actual world (see the planet in space), that is easy to do for the Copernican world, and hard and rather unpleasant to try to do in the (standard) Ptolemaic way (like trying to make yourself hit a tennis shot in an unfamiliar way).

16. Philosophers allowed that Ptolemy's system worked (it "saved the appearances"), but not that this elaborate apparatus in the sky could be real. This would be an impressive case for the virtues of philosophic contemplation, except that when the heliocentric system came along they were sure that it too must be false. It appears, consequently, that the philosophers were only reflecting everyday habits of mind, not some profound insight. But as I have already stressed [7.3], for astronomers, who worked in detail with this system, the epicycles were not *rebutted* by common sense, but were evidence of the modest powers of common sense compared to the power of mathematical analysis. As Plato warned, common sense yielded only shadows of reality, not the wonderful discoveries by which mathematics revealed the very structure of the world. But the Copernican system provided an explanation of all the appearances without the need to actually believe in epicycles.

What about the many epicyclets that remained until Kepler dispensed with all of the Ptolemaic apparatus? They could hardly have the same status as the major epicycles, since they could not be seen in the vivid way, and with the large consequences, that a trained astronomer could see the major epicycles. Once the major epicycles were gone, the credibility of the minor epicyclets as real objects in space (rather than as convenient devices for computation) was bound to fade, just as belief in the solid spheres was bound to fade for astronomers who had come to see the world in the Copernican way, where the spheres no longer play a physical role in determining distances in the universe (*Patterns* [13.6]). So it is not surprising that Kepler, early in his career while still using the epicyclets in his own work, simply waved aside as not worth worrying about the question of how to fit the epicyclets into his nested Platonic solids argument about the structure of the heavens.

Chapter 9

1. How can we know that? Because people who take the trouble to become familiar with the Ptolemaic models and acquires some experience working with them finds themselves thinking that way. But it is now no longer a significant barrier. Rather, as I will be discussing further, a modern person with Ptolemaic experience inescapably has entrenched Copernican habits of mind, and seeing things in the Ptolemaic way is something that needs to be "switched on." For a Ptolemaic astronomer there would have been no alternative way of seeing things: it need not be switched on, nor could it be conceived as being switched off, for there was no alternative.

2. I give some details on the Waldseemüller map and the very direct evidence that Copernicus had studied it closely in *Patterns* [12.9]. The map was forty-eight square feet (six feet by eight feet), printed in a dozen sheets, and

intended to be plastered on a wall. The edition was enormous for the time (1,000 copies), but only one has survived (bound as a book), presumably because the maps were, in fact, mounted on walls by owners anxious to see for themselves and show off to their friends the astonishing discovery.

3. In the Ptolemaic setup, the deferents of Mercury, Venus, and the Sun all have the same period. Since Ptolemaic astronomers never recognized the linkage motion, the periods of the epicycles of Mercury and Venus (144 and 584 days, respectively) seemed to bracket the period of the Sun (365 days).

4. The qualification "once primed . . . " is important. It is also worthwhile noticing that, even if the heliocentric possibility were on the table, this argument is not so striking against the inverted Ptolemaic possibility, where there is a direct physical link that accounts for coordination of motions.

5. This warning suggests a further mild indication (not mentioned in *Patterns*) that Copernicus indeed was interested in this issue. How would Osiander, who was not an astronomer, have known about this anomaly? Unless there is some evidence that the anomaly was common knowledge at the time, it would be a puzzling thing for him to know. But if in fact Copernicus was prompted by interest in just this anomaly, then it is not so surprising that Osiander, who was in contact with Copernicus and with Rheticus (whose task in seeing Copernicus's book through the press he completed), should have heard something of the matter.

6. Conceivably Kepler had some evidence. Certainly on another historical point (identification of Osiander as author of the preface to *De Revolutionibus,* where Kepler also mentions no evidence), independent evidence has been found confirming that he was offering more than an unsupported conjecture.

7. The sines Copernicus lists here give values of $r:R$ for each planet listed, since when the planet is seen at maximum elongation from the center of the epicycle, an observer on the Earth is looking at a right triangle with hypotenuse R and side r, as in figure 7.3. Of course, Ptolemy could not see the center of the epicycle. But he could compute its position and observe the variation of the observed planet from that computed position. Let x be the maximum value for the angular displacement on the epicycle. Then $\sin x = r/R$ will be the value Copernicus has extracted from the table of sines he had bound with his copy of the Alphonsine tables. But the value of r/R is identical before (in the Ptolemaic setup) and after (in the Copernican setup) the moves described in section 7.5. Let D be the Earth-Sun distance, which Copernicus sets as 25,000 units. Let D^* be the heliocentric distance of the planet in the same units. Then for the inner planets, where R is the s-orbit, the heliocentric proportion is $D^*/D = r/R$, so that $D^* = Dr/R = D(\sin x)$. For an outer planet, where r is the s-orbit, the proportion is $D^*/D = R/r$, so $D^* = DR/r = D/(\sin x)$.

8. A Ptolemaic astronomer would call the standard model an epicycle carried on a deferent, and the inverted model would be a deferent carried by an eccentric. Why is the number for Mercury also labelled *ecce*? Given that he has written *ecce* for the three outer planets which come first, that could be just a slip. This would be especially the case if Copernicus had been exploring inverted

models of the inner as well as the outer planets, though at this point only the inversion of the outer planets is needed. Alternatively, as Swerdlow suggests, perhaps Copernicus was in fact using an inverted model of the inner planets in working out the heliocentric system, which would also work [7.6].

9. That (perhaps later) the Moon was added to this list may reflect attention to checking that there was room between Venus and Mars for the orbit of the Moon (see *De Revolutionibus* I.10).

10. Ragep (1990), in the context of a critical discussion of Duhem, reviews the ample evidence that the Ptolemaic tradition, starting with Ptolemy himself, was always concerned with cosmology, not merely with "saving the appearances."

Chapter 10

1. And indeed in the Ptolemaic case something was really there that yielded observations *as if* epicycles themselves were real, just as something was really there that yielded observations *as if* Saturn's rings were real. But eventually someone noticed that a different assumption about what was there would yield the same observations in the epicycle case, but not so for the rings of Saturn.

2. Meaning here, of course, the major epicycles that account for retrogressions of the planets. See chapter 8, note 16.

3. But the appeal of a nesting principle did not so easily disappear. Kepler's first book was an exposition of how the Copernican planetary orbits fall within spaces formed by nesting the Platonic solids. Jardine (1984) provides a translation and commentary.

4. Copernicus's remarks here have often been read oddly, as either favoring a finite universe (though he seems emphatic in leaving the question open and bold to even raise such an issue), or as exhibiting the astronomer's expected modesty on questions where natural philosophers might claim superior competence. But the tone seems more inspired by Ptolemy's opening remarks in the *Almagest.* Ptolemy there is emphatic in ranking philosophers' talk about the cosmos as mere speculation in contrast to the reliable knowledge that can be provided by mathematical astronomy. Copernicus says nothing confident can be said about whether the universe is finite or infinite, so the question might as well be left to the philosophers. The remark seems to me to have more of a tone of superiority than of humility.

5. The secondary literature abounds with claims that the evidence provided by the new star of 1572 and the comet of 1577 were blows to Ptolemaic belief. But as discussed in *Patterns* [13.7] and in chapter 8, note 2, neither claim is plausible. Kepler at one point remarks that an argument from the comet provided by his teacher (Maestlin) helped persuade him that the Copernican view was right. But from his earliest book (1596), when he summarizes the main points of the Copernican case, no such argument is ever mentioned. There is no reason to doubt that Maestlin thought he had an argument here, or that as a young student Kepler found it helpful, or that an older Kepler was pleased to

have an occasion to tip his hat to his mentor. A person who is leaning toward some belief (for whatever reasons) sees arguments favoring that position in a kinder way than someone skeptical of where they are leading.

6. Consider some event that would ordinarily be analyzed as the interaction of two individuals, say, a championship match in tennis or chess. We expect that an account of how such events (world championship matches) come to *exist* must be largely sociological. An account of how some individuals within this community come to be so devoted to the special skills involved, and to achieve the levels of expertise they have, will also be largely sociological. No individual living in a community that did not cultivate these skills could perform as these championship contenders are able to perform. So there is a sense, and not a trivial sense, in which the tennis or chess match is a social construction. But ordinarily, it would clutter or distort what is going on in a particular match to insist on treating that particular match in sociological terms. The question at issue in the text here is how far a scientific controversy, as between Hobbes and Boyle or between geocentric and heliocentric astronomers, is likely to be clarified, rather than cluttered, by sociology. For example, the contagion phase (spread of a new theory across a community) obviously is a social process, like the spread of any other new idea. The issue here is whether local and contingent details of this process help us understand anything more than secondary details of why the particular idea took hold or failed to do so, or whether it is only some general properties of such social processes (not peculiar to the local situation) that yield a general constraint on the emergence or contagion of new theories.

7. As will be seen, this is particularly unavoidable in dealing with the influence of evidence. Knowing some things about the way the world is lets us know a good deal about what observational evidence could be robustly available, and what could only be artifactual if reported at all. In general, there is a bootstrapping problem here. Unless there is something we can say we know about the world, how could we say anything sharp about the possibility that what we believe is shaped by the way the world is? But if we suppose there are things we know, we have already pretty well assumed what we are trying to show. The explanation involves a retreat to a coherence account of truth. In a recent study, Torretti (1990, 7) comments, "In the trial of empirical knowledge the defendants are at once the prosecution, the witnesses, and the jury, who must find the guilty among themselves with no more evidence than they can all jointly put together."

8. The focus here is on *scientific* belief. For beliefs in general, including lay beliefs about science itself, it is hard to imagine anything sufficiently bizarre that we could be confident no one believed it. Defining *science* is famously difficult, though not quite *so* difficult as defining *love, good, true,* and various other terms no one finds vacuous, however hard it is to pin them down in a formal definition.

9. There happen to be what might be claimed as counterexamples, such as the front legs of a fiddler crab. But—and just the point of the argument here—the legs are no longer used for walking.

10. I distinguish propensities from habits, since it seems useful and

perhaps is significant at the level of how the brain works, that some things prompt actions (understanding that actions may be wholly internal, like formation of a mental image), but others are simply evaluative: something looks right or looks wrong. In the discussion here, I usually mean the former when I say habits, the latter when I say propensities.

11. Science itself provides ready examples of beliefs that logically seem to be little more than a matter of convenience, since the very individuals holding the beliefs concede that the evidence is not really sufficient to choose between their favored position and some available alternative. The attitude of Lavoisier and most other scientists in the decades before and after 1800 with respect to the nature of heat is a good example (chapter 4, note 5).

12. As usual, this needs to be understood as a remark about propensities not consciously perceived by the chooser. The chooser simply discovers that her mind is made up.

13. The situation in science is not very different from what is going on almost all the time outside the context (mainly mathematical) of formal argument. A writer or speaker is typically making a claim, presented in a way she expects to prompt an intuition of plausibility in a hearer or reader. Sometimes that is disappointed, and the resulting failure of intuitions to cohere (so controversy or misunderstanding arises) has a salience that routine agreement does not have. But routine, unnoticed coherence of intuitions is what is going on in the great majority of individual instances. When it isn't, we come to feel uncomfortable and decide we "just can't talk to" that person.

14. The case of modern physics versus Newtonian physics is not essentially different from the Copernican/Ptolemaic case: the nested-spheres presumption of Ptolemaic astronomy dealt with things *beyond* observation, as did the presumption that Newtonian mechanics worked for phenomena moving close to the speed of light or on a subatomic or extragalactic scale. It makes little sense to say that pre-Copernican astronomers or prequantum physicists erred by assuming things that couldn't be observed. As I have been urging, pursuing conjectures can't be considered an error if, realistically, that is how the human brain does its work. Cognitively, commitments—on the argument of this study, entrenched habits of mind that amount to implicit commitments—can't be avoided. Hence the importance of Kuhn's (1977) "essential tension." If we were logical automatons, this tension would involve only a narrow point about the allocation of resources across possible avenues of research. Serious issues would rarely arise, since a sensible person would rarely be moved to put all his eggs in one basket. But in fact we all make far more one-sided commitments than are logically warranted. That is not something we choose to do, or that we ordinarily can notice, but just the way our brains work. Further, such (occasional) excessive commitments are not helpfully construed as creating an interest in any interesting sense of the word *interest*. To call a habit of mind an "interest"—even a "cognitive interest"—seems to me like saying that a drunk has a physiological interest in driving erratically, or that a weak student has a muddling interest in doing badly on his exams.

15. I mean here, of course, cases in which evidence has been accepted

as uncontroversially correct. Novel evidence is often doubted, and such controversies often end (as with recent episodes involving cold fusion, polywater, and gravitational waves) with a consensus that the evidence was artifactual.

16. What about the various homocentric schemes, from Aristotle's times right down to Copernicus's? It seems to me they *illustrate,* not challenge, the main point I'm making. None of these schemes managed to match observations nearly as well as the Ptolemaic. Although renewed efforts of this sort appear over and over, nothing like a long-lived school of homocentric astronomy ever developed. Since it is easy to point to credible interests that would favor such a scheme (philosophers and theologians over and over say that that is how the world *ought* to be, hence how it must be), and since the social standing of the philosophers and theologians was always higher than that of the mathematical astronomers, on an interests view of science it is a real puzzle (I should think) why homocentric astronomy never got to be anything more than a quirky, intermittent sideshow in the story of astronomy. But if in fact the world constrains mathematical astronomy to theories that fit an orbiting Earth (which we have seen that the Tychonic and Ptolemaic, as well as the Copernican, theories do), then obviously the homocentric schemes couldn't compete, *even though* backed by socially powerful supporters with a clear commitment to their success.

17. That the Copernican triumph was not driven by the sheer rigor of the Copernican case is indicated by how popular the view remains among historians of science that firm Copernican convictions were premature at least until Newton's *Principia.* See the comments in Margolis 1991, note 24.

18. Perhaps Gilbert should be added here, as discussed in chapter 12.

19. Again I am ignoring here various secondary epicyclets used by Ptolemy and Copernicus.

20. Why isn't the situation like that of a Rorschach test, where there is no limit on the number of different ways someone might see a blot? Because the ink blot is just a blot, indeed, a blot deliberately chosen for its susceptibility to many interpretations. There is nothing that the ink blot "really is." Since there is nothing really there to impose coherence, a person looking for a pattern eventually finds some recognizable piece of the blot to take as figure, leaving the rest as ground. Since there are indefinitely many ways that might be done, there are indefinitely many interpretations, contingent on individual propensities and background experience (patterns already in the individual repertoire). But for the planets, no observer can miss the primary cycle through the zodiac and the secondary cycles that coincide with oppositions to the sun.

Chapter 11

1. Page references are to Shapin and Schaffer 1985. S&S refers to their text; H refers to Schaffer's translation of Hobbes's *Dialogus Physicus,* printed as an appendix to the book.

2. Why would Hobbes proceed this way? A possible answer is suggested in the way he handled his change of heart about the role of particles.

Instead of changing the passage (discussed further below) in which he asserts an explanation in terms of the particles, Hobbes simply tacks on a passage pronouncing that claim incredible at the very end of his dialogue (H 391). This is separated from the discussion of Boyle's experiments by eight pages of one of Hobbes's favorite geometry topics (duplication of the cube). So on the face of things, Hobbes was not inclined to revise what he had already written. But the particles argument, as he uses it, looks like it was originally devised to deal with Pascal's Puy de Dôme experiment many years earlier. So Hobbes perhaps started out with some material that predated Boyle's experiments entirely. When he came to deal with Boyle's work, Hobbes switched to his "vehement wind" hypothesis but never bothered to revise the earlier pages. But the bottom line here is absolutely clear. Rather than "again and again" explaining Boyle's pressure effects as due to the motion of particles, as Shapin and Schaffer assert, Hobbes emphatically denounces that as a mistake (H 391). Instead, again and again—in the dialogue itself and in several summaries of his views published later—he attributes all the Boylean effects to the vehement wind.

3. Shapin and Schaffer give credit to (or put blame on) Boyle and the politics of the English Restoration for the commitment to publicly witnessed experiment and distrust of speculation not clearly tied to observation. But there seems to be little in this that could not be more realistically tied to Pascal's work a decade earlier. In addition to the famous Puy de Dôme experiment described here, this included Pascal's replication of the Torricelli experiment before a large audience using a tube over thirty feet long filled with wine to show that, contrary to his adversaries claims, wine (being lighter than water, even though more volatile) would stand higher than water. Was this only a demonstration, not an experiment? That sort of claim will be taken up in chapter 12.

4. Hobbes (H 382) was confident that water cannot expand when it freezes.

5. Hobbes once or twice uses the Cartesian term "ether" for the space-filling fluid., and Boyle complains he is not sure just what Hobbes wants to claim. But very little of what Hobbes has to say is coherent unless by "ether" or "pure air" (another rarely used term) he means ordinary air, less particles of dust. Hobbes's pure air (or ether) is a substance that can blow apart a bladder, smother animals, and distend the walls of Boyle's pump. It is not Descartes subtle ether than can pass through the pores of ordinary material.

6. Sometimes Hobbes speaks of air being compressed, but in contexts where no smaller volume is being supposed: compressed seems to mean something like "pressed upon." I have not noticed any place in the *Dialogus* in which Hobbes explicitly says that he takes air to be incompressible in our usual sense, but he makes the more general remark (H 384) that a given amount of a body always fills the same space, so that ice cannot actually expand when frozen, but must absorb a volume of air equal to the apparent increase in volume. In unpublished notes, as Schaffer himself has reported, Hobbes is explicit on the point.

In any case, either Hobbes meant that a given amount of air has a strictly fixed volume, or his argument here and at many other points is wrong on its face. For there would be no point to Hobbes's claim that the air that fills the universe could not accommodate the tiny volume equivalent to the empty space at the top of the Torricellian tube unless it is strictly incompressible.

7. Hobbes's criticism of the air pump is often framed as a claim that it leaked. But that was never the issue. Boyle himself allowed that the pump leaked, and Hobbes explicitly notes that he was aware of that (H 366). Hobbes's claim is that the pump leaks like a sieve: as fast as air goes out it slips back in. Further, that was not from some imperfection in Boyle's design, but because in a full world an air pump was logically impossible. If the seal around the piston were completely airtight, the pump would not operate, since it would then be impossible to move the piston.

8. Since Hobbes's primary explanation of the Torricelli effect turns on the weight of mercury required to press air through fluid (with the particles playing only a secondary role in accounting for why the effect is greater on top of a mountain), his account gives no explanation of why the empty space disappears if the tube is tilted. Any air in the apparently empty top of the tube can't be pushed down by the weight of mercury, since on Hobbes's account that is still directed down, not against the space above it.

9. Shapin and Schaffer discuss this exchange and quote the remark just given, but only as part of a discussion of the rhetorical structure of the *Dialogus Physicus* (S&S 144–45). They do not mention the substantive point of the exchange, and they also ignore the substantive point in their account of Hobbes's argument, reporting the particles explanation (S&S 122) with no indication that it was later emphatically withdrawn.

10. The list can be extended. If air has a fixed volume, where is that volume while the air passes invisibly through water and mercury? Why does a column-inch of Hobbesian air rush invisibly down through mercury if we pour an additional inch of mercury in the basin? There is no force pushing it out in Hobbes's account, and it has to go two inches further through the mercury to get out. If the vortex provides a powerful outward force to rush out and push up the piston, how can it also provide a powerful inward force that prevents a lid on the receiver from being lifted? And so on.

11. Boyle's reply to Hobbes, introduction.

12. For example, in a passage from Hobbes that Shapin and Schaffer partly quote (see note 13), after a remark by the interlocutor praising experimenters, the spokesman for Hobbes allows that indeed they enrich the fund of information without which true knowledge of nature would be sought in vain.

13. Here are two examples. Shapin and Schaffer quote Hobbes as saying of Boyle and his colleagues that "all of them are my enemies," asserting that this shows Hobbes's view of the experimental program (S&S 112n). But in context (H 347–48), the "all my enemies" refers to politicians, clergy, and mathematicians in addition to his Royal Society (Gresham College) adversaries. The following lines, which say why all these people are his enemies, deal exclusively with examples of personal abuse Hobbes says he has suffered. In no case do the examples have anything to do with experiment.

A few pages later (S&S 128), Shapin and Schaffer translate a Latin remark of Hobbes's as saying that "if the sciences were said to be experiments of natural things, then the best of all physicists are quacks." They give no argument for translating "pharmacopoei" here as "quacks," though indeed the term could carry that connotation. But in the context at hand that seems unlikely. Hobbes is emphatic in allowing that experiments are the raw material for scientific discovery, but also in insisting that someone who merely marvels at the results of experiments is like a child who admires a book but does not understand its text. If all there is to science is gathering up phenomena, then compilers of pharmacopoeia would be the best scientists. But on Shapin and Schaffer's reading, Hobbes is not saying that merely experimenting does not make a good scientist. They turn that around and read him as claiming that doing experiments makes a bad scientist. Shapin and Schaffer soon harden Hobbes's remark about "pharmacopoei" into a claim that "adopting an experimental form of life changed physicists into 'quacks'" (S&S 129) and then that "for Hobbes, the best experimenters were quacks" (S&S 307). We see an escalation of a remark that in context Boyle could almost as easily have made as Hobbes into a bald claim that for Hobbes experimenters are quacks. That is entirely characteristic of the rhetorical style of the book. Here is the full passage (A speaks for Hobbes, B is the interlocuter), translated by Noel Swerdlow.

> A. For the natural cause of every thing is some motion. But those who are now applying themselves to philosophy most of all have considered the nature of motion least of all; on this one thing they dwell, that they may obtain new phenomena, since only by learning of new phenomena by experience [*experiendo*] are causes to be learned by reasoning [*ratiocinando*] from motion.
>
> B. Those who see new and marvelous works of nature by applying bodies to bodies wonderfully inflame the minds of men with love of philosophy, and in no small way incite them to investigate causes, and for that reason are worthy of praise.
>
> A. That is true for they enrich natural research [*historia*] without which natural knowledge [*Scientia*] is sought in vain. But to gaze upon and admire the works of nature, as a child looks upon the beauty of a book more than the [meaning of the] writing, does not belong to a philosophical man. That is what those do who, seeing phenomena, do not consider by what agent, by what motion, and by what means they could have been produced. For if experiments are worthy to be called knowledge of nature [*experimenta rerum naturalum scientia dicenda sint*], the best natural philosophers of all are the compilers of pharmacopoeia.

14. Hobbes (like Boyle) was a Copernican and enormously admired Galileo. If Shapin and Schaffer were correct in their claim, Hobbes would presumably have approved of the church's exercise of its sovereign responsibility to suppress Galileo's threat to order within its jurisdiction. And he would have found it reasonable for scientists in France to be bound to be vacuists, in England bound to be plenists, or whatever else their respective sovereigns favored. Hobbes's Leviathan argument with respect to politics and religion is well known, but Shapin and Schaffer's extension of this argument to science seems to be their

own invention, turning on a claim that, if Hobbes is read sufficiently literally, then belief in the possibility of a vacuum threatens to undermine the state.

15. I don't mean to imply that there is no difference among the scientific styles of Boyle, Pascal, and so on. But all, both in the way they acted and the way they talked about science, fell within the range of modern science. As I have indicated, I am not sure that Hobbes himself clearly falls outside that range, although he is certainly on the edge of doing so. The views that Shapin and Schaffer *attribute* to Hobbes (S&S 119–20) are another matter.

16. On Shapin and Schaffer's view, the outcome reveals who "has the most, and the most powerful, allies" (S&S 342). Indeed, the Boylean way in science is more consistent with the limited monarchy of the Restoration than the absolutist position on how science should work that they attribute to Hobbes. Yet Hobbes's attack on Boyle was no more successful outside England than in it, so that the tie to English politics Shapin and Schaffer stress seems no more than an especially flagrant example of *post quo, ergo quo.* Shapin and Schaffer do not document any substantial difference between experimental science as practiced in England circa 1660 and its analogue on the Continent. Hobbes's friend Sorbière annoyed his English hosts by writing what they took to be an insufficiently appreciative account of English life and society. But contrary to the impression given by Shapin and Schaffer (S&S 112n), Sorbière's (1664, 31–38) appraisal of the Royal Society and how it conducted its meetings was on the whole extremely complimentary. Similarly, there is no striking difference, so far as I can see, between Boyle's view of science and that expounded by Pascal in France a decade earlier. Leavenworth (1930) gives a detailed exposition. For a contrary argument, see Dear 1990.

17. Redondi (1987) tries to show that the pope was forced to act against Galileo because of his heretical views about the atomic nature of matter, and the Copernican focus of his show trial was merely to cover up the more scandalous episode.

Chapter 12

1. Particularly instructive is Schmitt's (1951) discussion of Harvey's account of his methodology. The sense in which Harvey proclaims himself Aristotelian is just the sense in which (Schmitt points out) Galileo can construe himself a follower of Aristotle.

2. For a general discussion, see Hooykaas 1987.

3. An early candidate for a clearly contrived experiment—something not in some obvious way related to ordinary experience—is provided by Schmitt (1969) where a "syphon" (we would call it a straw) is used to suck air out of a glass ball. This is then inverted in water, straw down, and water is seen to be sucked up into the ball. This is one of a number of experiments discussed in Schmitt's survey of sixteenth-century debate over the possible existence of a vacuum. Clearly we have a device here specifically contrived for the occasion (in contrast to other experiments, which involve only familiar objects, like bellows). But it is also easy to conjecture how this device came into being,

prompted by debate about what happens when a person sucks fluid through a straw into the mouth. By circa 1640 not only are there many such contrivances in use—not merely one obscure example from a century—but the contrivances are no longer linked in some immediately obvious way to uncontrived experience.

4. Perhaps *indirect* would be a better label, since there are contexts of observation that are not contrived, but nevertheless are strikingly indirect. The best example is the remarkable Copernican argument Galileo extracts from observations of sunspots. Who would suppose that keeping close track of the appearance of blemishes on the surface of the Sun could yield a powerful argument for the rotation of the Earth? But perhaps even Galileo himself would not have seen that, unless already habitually alert to the possibility of evidence in contexts away from the immediate focus of inquiry.

5. Note especially that in the 1650s Guericke in Germany and Sorbière and colleagues in France as well as Boyle in England were pursuing the conceptually simple but pragmatically tricky effort to build a workable air pump. There was nothing peculiarly Boylean, or English, in seeing that as the way to carry the intensely discussed work of Torricelli et al. into its next phase. Boyle, with Hooke's help, got there first. But that someone would get there soon was beyond reasonable doubt, as was the inevitability that the striking results that came with a workable air pump would attract attention throughout Europe. What could be seen, once a workable air pump was available—whether written up in Boyle's modest style, or Galileo's immodest style, or Newton's difficult style, or Pascal's comfortable style—was unambiguously an exciting next episode in a story that had already commanded the attention of everyone interested in the new, experimental philosophy.

What about opposition to this development? As far as I can see there was none of importance. I will say something later about why that was so. I have already commented [11.8] on Shapin and Schaffer's claim that Hobbes was attacking experiment as a central feature of science.

6. So I will try to suggest how an emergent habit of mind is the crucial development, and I will give a special salience to Gilbert's work on the magnet, though (as in the case of probability) that emergence is on the doorstep, so to speak, and it is hard to believe it would be long before someone opened the door and noticed what was there. The shift is irreversibly put in place before 1650, as I have mentioned, probably with the publication of Galileo's books in the 1630s. It is only a conjecture that Galileo might not have played quite so decisive a role if not stimulated by Gilbert's much-admired work in which a molded lodestone (the *terrella*—a kind of toy Earth) is put to work to produce both striking practical results (about the behavior of mariners' compasses) and also an account of the Earth's rotation that we now know was mistaken but that greatly impressed Copernicans and anti-Copernicans alike. In any case, in Galileo's work contrived observation is unproblematically part of natural philosophy. When later writers self-consciously speak of the "new philosophy," what is most explicitly new is the prominent role played by appeal to contrived observation. The "new philosophy" is synonymous with "experimental philosophy." Hobbes spoke for many when (contrary to what might be expected on Shapin

and Schaffer's argument) he called Galileo the first to pursue natural philosophy in a proper way.

7. One common claim about what is novel about experiment in the seventeenth century seems to me definitely wrong. This turns on a claimed deep contrast between experiment as demonstration and experiment as test. It is certainly the case that an investigator (a theorist looking for an experiment that fits a theory, a mathematician looking for a proof that fits a conjecture) is sometimes in a frame of mind where she feels subjectively sure of her result, and the new evidence is merely a demonstration to convince other people of what she already thinks she knows. But cranks aside, the same person also has doubts (though not at the same moment, for these are two genuinely distinct gestalts), so that success in finding a replicable experiment or a publishable proof becomes a test of whether the idea being pursued really warrants belief. Similarly, the rhetorical situation at hand might sometimes encourage a person to present results aggressively, as a demonstration of what careful reasoning on prior evidence would affirm, though if rhetorical circumstances were different the same results might be presented modestly, as a test of an idea that might very reasonably be doubted.

So psychologically and rhetorically, there are distinctions to be made between experiment as demonstration and experiment as test. But if a demonstration comes to be expected, then a failure to provide one becomes an argument that something is wrong with the claim. The ability to provide a demonstration then becomes a test. Conflicting demonstrations demand a resolution. Neither claim will be highly credible until one has been crippled.

Galileo rhetorically suggests (in the *Dialogue,* 141ff., but see the interestingly different treatment in his 1623 reply to Ingoli [Finocchiaro 1989, 184]) that pure reason assures that a stone will fall straight down from the mast of a moving ship, not to the rear as his adversaries expected. So the rhetoric here treats the experiment as merely a demonstration. But this rhetoric blatantly challenges his adversaries—and probably more important for Galileo's strategy, invites parties undecided about his Copernican argument—to see for themselves. Since Galileo will look ridiculous if this easily tested claim does not prove true, it will certainly function as a test of his argument. As is true in general, a practical line between experiment as demonstration and experiment as test cannot really be sustained.

8. The usual interpretation is that Gilbert was silent on the annual orbit. But there are a number of remarks in his discussion (1600, concluding chapter) that are at least odd if he does not wish his readers to understand his position as fully Copernican. He says daily rotation is the "only motion for which I will spell out the reasons here," which, in the absence of some hint—of which there is none—that he is in doubt about the annual motion, suggests that if the occasion called for it (that is, if he could tie it to his account of magnetism), he would provide reasons for believing the annual motion as well. A few paragraphs later, after a comment on the Earth, he refers to similar effects for "the other planets." A person who even so indirectly labels the Earth a planet is, at the least, not much concerned about being mistaken for a Copernican.

9. The response to Galileo's telescope is not fully germane here, since

the telescope involved contrived means of observation, not a contrived context of observation. But the objections at first raised to Galileo turned almost entirely on doubts that the instrument was reliable, or that Galileo himself was reliable, not noticeably on any question of principle about contrived observations. On Hobbes, see sections 11.8 and 11.9.

Chapter 13

1. Suppose partisans on both sides accept enough in common that a rigorous proof of advantage to one side is available. But even when this seems so, it might fade. Confronted with arguments that make a position seem untenable, a partisan might conclude only that some premise or piece of reasoning taken as uncontroversial really needs to be questioned.

2. Understand internal factors to be the closely worked-out interweaving of argument and evidence that we see in texts. External factors are then whatever else might have influenced belief, by influencing the structure of or the response to the internal factors, but with habits of mind a distinct category (see note 3).

Naturally, complications arise. What historical actors took to be internal may not coincide with later ideas, the usual example being appeals to the authority of the Bible in the debates over Copernicanism, geology, and Darwinism. And arguments do not come nicely labelled as to what is rhetorical flourish and what is internal argument. But so far as I have seen, rough but adequate distinctions can be made in concrete cases between what is part of the close reasoning pushed up against close observation that characterizes internal factors, and what isn't.

If we think of internal factors as argument and evidence that we could expect to appear in a textbook, then habits of mind are obviously not internal factors. A person might explain an intuition as due to a habit of mind, but (illustrating the point) most often when he has come to see the intuition as mistaken. No one tries to justify a belief on the grounds that it suits his habits of mind, although a person might explain a practice that way. But habits of mind are not external either, in any sense I've found useful. So I treat them apart from any internal/external distinction.

Endemic propensities are similar, even though appeals to economy and comfort appear routinely as explicit grounds for finding some arguments plausible and others not so. These propensities are operating whether appealed to or not, occasionally with consequences that would be seen as unreasonable if recognized.

Why should not the point apply as much to the habits-of-mind influences? Because habits of mind are tuned to individual experience, so that as evidence accumulates and familiarity with arguments grows, even a very stubborn barrier is likely to be overcome eventually, provided there is something stubbornly out there that resists some ideas and accommodates others. The underlying commitment of the habits-of-mind analysis to the tuning of habits to experience is incompatible with relativism. Rather, as will be discussed in more detail later, we are alerted to especially striking roles of habits of mind by notic-ing when scientific beliefs that prove to yield powerfully effective accounts of how things go in the world are either remarkably slow in emerging, or remark-ably rapid. The long-delayed emergence of probability and then the rapid exten-sion of stochastic to epistemic probability illustrate the polar cases [6.12].

3. Another example has come up among the cases presented here: the preference (circa 1800) for the caloric theory of heat over the mechanical, by nearly everyone other than Rumford (chapter 4, notes 5 and 10).

4. But then aren't the people for whom the black box is an open book just those whose expectations are solidly shaped by commitment to the theory? Not entirely. Once a piece of theory is routinely covered in texts, there will be many people around who know enough to "get up to speed" if a puzzle seizes their attention. Here the individual does his tuning up specifically focused on an anomaly. More important, the propensity to miss or dismiss as somehow mis-taken evidence that conflicts with expectations is not absolute. Here some clear anomaly has emerged—it is no longer something that no one has noticed. Of the many people who know the theory, only one who takes the anomaly seri-ously may be required for it to be sharpened to the point where it becomes impossible for others to ignore.

5. Compare the situation of code-breakers trying to decipher an ene-my's secrets to that of scientists trying to decipher nature's secrets. For one context as much as the other, the technical work must proceed without an op-portunity to peek behind the curtain and see what is *really* there. Code-breakers are typically scientifically trained people going about their business in very much the way they have learned to deal with scientific problems. Modern code-breaking characteristically involves team efforts, using computers and other black-boxed technology and technique. Work proceeds with a lot of talking through of things (what has come, oddly, considering ordinary language con-notations of the term, to be labelled *negotiating*), with gradual construction of a consensus. In all, code-breaking seems to be the product of social construction in every sense that science is.

For code-breaking, difficulty is enormously eased because the code-breakers ordinarily know enough about the lexicon and grammar of the mes-sages likely to be transmitted to have strong grounds for recognizing a good solution. On the other hand, difficulty is aggravated compared to the case of science, since nature presumably is not trying to hide its secrets, while code-makers are trying their best for that.

This code-breaking analogy gives us a concrete example of a process of sciencelike social construction that in fact yields products that meet a very reasonable standard of rational construction. We have a real case that in quite a detailed way is like the case of science. But while we don't have access to meta-physical truth, and hence could never know whether we are in some reasonable sense getting nearer to it, in the parallel case of code-breaking we do get to see the right answers (at least for those who lose the war).

6. Cases in which this seems to be violated are often illusory. In one context a model that efficiently gives a good empirical fit is used, while in an-

other what seems a more realistic, richer, but analytically more difficult-to-apply model is used (statistical mechanics versus thermodynamics, or micro- versus macroeconomics). Most commonly in such cases, there is an overlap between the people using one theory and those using the other, responding to different contexts within a shared sense of empirical constraints. There is never a total split into noninteracting, independently evolving worlds. See the related comments in *Patterns* [7.2].

7. Shapin and Schaffer (chapter 6) show great concern about subtle discrepancies between different experimental results (such as Huygens's discovery of anomalous suspension), as if only *exact* replication by *exactly* the same experimental setup could count as evidence for Boyle's views against Hobbes. But as will be discussed, that has exactly the wrong intuition behind it. To the extent that some phenomenon appears only if we use *exactly* some experimental setup, we have strong reason to suspect it is an artifact. If it is real, there should be many ways to elicit the phenomenon.

Anomalous suspension could only be observed when pressure had been reduced to about 1% of ambient air, and then only when meticulous care was taken to elicit it. Hence a precondition for observing anomalous suspension, and for recognizing it as anomalous, is that Boyle's claims had already been effectively replicated. Shapin and Schaffer somehow see this discovery of an unexpected effect that could be elicited only by delicate use of the air pump technology as something that could have undermined confidence in giving a central role to experiments in science.

8. Feyerabend argues that Galileo won by outmaneuvering his critics, since he could neither account for inconsistencies in what the telescope showed, nor give theoretical reasons for supposing that the telescope would be accurate with respect to celestial objects. The major inconsistency Feyerabend stresses was that the edge of the Moon appears smooth even though Galileo claims to see high mountains away from the edges. But near the terminator, long shadows exaggerate the height of mountains. And at the edge, as Galileo argued, not only are there no shadows to help reveal contours, but we see mountains only by looking past or against the background of other mountains near them. But Feyerabend claims that photos from space satellites show that the Moon's mountains do not have the right configuration to support Galileo, which is at once anachronistic and false. Anything goes.

On the other hand, to objections that we have no reason to trust the telescope, Galileo pointed to cases where we know what the telescope should show if it is a reliable instrument. For example, Ptolemaic, Tychonic, and Copernican astronomies all agree that Mars and Venus should each appear about six times as large at opposition as at conjunction. With the naked eye Mars does not appear to vary by nearly this much, and Venus scarcely seems to vary at all. But for both, the telescope shows the predicted increase in apparent diameter. The moons of Jupiter show the successively slower periods that are compatible with the variation of planetary motions with distance from the Sun. And the phases of Venus reveal the specific form that makes sense if either Copernican or Tychonic astronomy is right. In all this and other matters as well, we have the remarkable coincidence that, where we have strong reason in advance to expect something that cannot be seen with the naked eye, the telescope just happens to show what it should show.

9. Maestlin (Kepler's teacher) had shown two decades earlier how the analogous measurement for a comet could be made using nothing more than a piece of thread to check motion over the course of a night relative to nearby fixed stars. An astronomer with Ptolemaic convictions did not need access to Tycho's instruments to challenge his claim. Nor was the observation one that could not be made for lack of opportunity. An appropriate time occurs at each opposition of Mars (that is, every two years). Nor was it the case that Tycho's prestige was so enormous that no one would dare challenge him: in fact he was challenged on his claims about the comet of 1777, though here we know he was right on his main claim (that the comet exhibited no observable parallax.) Maestlin had no reason to challenge Tycho: he was already a Copernican, and hence believed what Tycho claimed to show. The point here is that apparently every other astronomer with the competence to challenge Tycho was also ready to believe Ptolemy was wrong, as a number of other indicators also suggest (Margolis 1991). For an account more sympathetic to Tycho, see Thoren 1990.

10. See the discussion of logic-p versus logic-t in *Patterns,* chapter 5. Of course, our economist is trained to worry about overt contradiction, while the Azande respondent is not. If the contradiction becomes explicit, our economist will think of reasons why it is only apparent (self-interest is a good enough approximation in this context, and so on). Sometimes the reasons are in fact reasonable, but sometimes—in this case, as in others, including beyond any doubt others in which you and I would be doing the explaining—they are only papering over what is really an unresolved contradiction.

Appendix B

1. In a footnote, Feyerabend (1975, 126n) allows that in fact Galileo's observations were confirmed by impeccably orthodox observers before the year was out. He explains this as showing a failure to follow proper method.

2. Note again that the claim is not that this inference could not be wrong. We have reviewed in detail the Ptolemaic case where just such a fit turned out wrong, since it turns out that another, cosmologically radically different scheme also fits. But until someone can show us such an alternative, the natural response is that what fits remarkably well is believable. Nevertheless, even for Ptolemaic belief, as discussed in detail earlier, the Ptolemaic analysis was wrong in the cosmology it implied, but brilliantly right in the formal relations it posited. We continue to believe that two-orbit accounts of planetary motions are right, that the motions of the two orbits are independent, that the variations in distances of the planets from the Earth are just those that Ptolemy saw.

References

Abu-Mustafa, Y. S., and Psaltis, D. 1987. Optical neural computers. *Scientific American* 256, no. 3 : 88–95.

Bernstein, N. A. 1932. The coordination and regulation of movements. In H. T. A. Whiting, ed., *Human Motor Actions: Bernstein Reassessed*. Elsevier, 1984.

Boyle, R. 1972. *Works*. Coronet Books.

Branigan, A. 1981. *The Social Basis of Scientific Discoveries*. Cambridge University Press.

Bruner, J. S. 1957. Going beyond the information given. In H. Gulber et al., *Contemporary Approaches to Cognition*. Harvard University Press.

Butterfield, H. 1965. *Origins of Modern Science*. Free Press.

Cardano, G. 1662. *On Games of Chance*. Johnson Reprints, 1967. Trans. S. H. Gould in Ore 1953.

Cohen, I. B. 1985. *Revolution in Science*. Harvard University Press.

Collins, H. 1981. Knowledge and controversy. *Social Studies of Science* 11 : 1–21.

———. 1985. *Changing Order*. Sage Publications.

Conant, J. 1950. *The Overthrow of the Phlogiston Theory*. Harvard Case Studies in Experimental Science. Harvard University Press.

Copernicus, N. 1543. *De Revolutionibus Orbium Caelestium*. Trans. A. M. Duncan. Barnes & Noble, 1976.

David, F. N. 1962. *Games, Gods, and Gambling*. Hafner Publishing.

Dear, P. 1990. Miracles, experiments, and the ordinary course of nature. *Isis* 81 : 663–683.

Dennett, D. 1975. Why the law of effect will not go away. *Journal of the Theory of Social Behavior* 3 : 169–187.

Donovan, A. 1988. Lavoisier and the origins of modern chemistry. In *The Chemical Revolution: Essays in Reinterpretation.* Osiris.

Drake, S., and O'Malley, C. D. 1960. *Controversy on the Comets of 1618.* University of Pennsylvania Press.

Dreyfus, H. L., and Dreyfus, S. E. 1986. *Mind over Machine.* Free Press.

Duhem, P. 1985. *Medieval Cosmology.* University of Chicago Press.

Edelman, G. M. 1985. Neural darwinism. In M. Shafto, *How We Know.* Harper & Row.

Evans, J. 1991. "Saving the phenomena" in ancient Greek astronomy. Paper presented at the 1991 History of Science meeting, Madison, Wisconsin.

Evans-Pritchard, E. E. 1976. *Witchcraft, Oracles, and Magic among the Azande.* Oxford University Press.

Feyerabend, P. 1975. *Against Method.* NLB London.

Finocchiaro, M. 1989. *The Galileo Affair.* University of California Press.

Fodor, J. 1975. *The Language of Thought.* Crowell.

Galileo, G. 1610. *Siderius Nuncius.* Trans. A. Van Helden. University of Chicago Press, 1988.

———. 1623. Reply to Ingoli. In M. Finocchiaro, *The Galileo Affair.* University of California Press, 1989.

———. 1638. *Two New Sciences.* Trans. S. Drake. University of Wisconsin Press, 1974.

Galison, P. 1988. *How Experiments End.* University of Chicago Press.

Garber, D., and Zabell, S. 1979. On the emergence of probability. *Archive for History of Exact Sciences* 21 : 33–53.

Geertz, C. 1973. Person, time, and conduct in Bali. In *Interpretation of Cultures.* Basic Books.

Gilbert, W. 1600. *De Magnete.* Trans. S. P. Thompson. Basic Books, 1958.

Gingerich, O., and Westman, R. 1989. *The Wittich Connection.* American Philosophical Society.

Goldstein, B. R. 1967. The Arabic version of Ptolemy's planetary hypothesis. *Transactions of the American Philosophical Society* 57, part 4.

Griggs, R. 1989. To "see" or not to "see." *Quarterly Journal of Experimental Psychology* 41A : 517–29.

———. 1990. Instructional effects on responses in Wason's selection task. *British Journal of Psychology* 81 : 197–204.

Hacking, I. 1975. *The Emergence of Probability.* Cambridge University Press.

———. 1986. Review of J. B. Cohen, *Revolution in Science. New York Review of Books* 33 (February 27): 21–23.

———. 1991. Artificial phenomena (review of Shapin and Schaffer 1985). *British Journal for the History of Science* 24 : 235–41.

Harvey, W. 1651. Introduction to *On Generation.*

Heilbron, J. 1990. Book review. *Historical Studies in the Physical Sciences* 22 : 198.

Henle, M. 1962. On the relation between logic and thinking. *Psychological Review* 69 : 366–78.

Herrnstein, R. J.; Loveland, D. H.; and Cable, C. 1976. Natural concepts in pigeons. *Journal of Experimental Psychology: Animal Behavior Processes* 2 : 285–311.

Hobbes, T. 1661. *Dialogus Physicus.* Trans. S. Schaffer in Shapin and Schaffer 1985.

———. 1662. *Seven Philosophical Problems and Two Propositions in Geometry.* Reprinted in vol. 7 of his *English Works.* Scientia Verlag, 1966.

———. 1966. *English Works.* Scientia.

Holland, J. H.; Holyoak, K. J.; Nisbett, R. E.; and Thagard, P. R. 1986. *Induction.* MIT Press.

Holmes, F. L. 1985. *Lavoisier and the Chemistry of Life.* University of Wisconsin Press.

Hooykaas, R. 1987. The rise of modern science: Where and why. *British Journal for History of Science* 20 : 453–73.

Hull, D. 1988. *Science as a Process.* University of Chicago Press.

Huygens, C. 1657. *The Value of Chances in Games of Fortune.* Trans. W. Browne, 1715.

James, W. 1890. *The Principles of Psychology.* Dover, 1950.

Jardine, N. 1984. *The Birth of History and Philosophy of Science.* Cambridge University Press.

Kirwan, R. [1784] 1789. *An Essay on Phlogiston.* With comments by Lavoisier et al. and reply by Kirwan. Cass Reprint, 1968.

Kitcher, P. 1993. *The Advancement of Science.* Oxford University Press.

Kuhn, T. [1962] 1970. *The Structure of Scientific Revolutions.* 2d edition. University of Chicago Press.

———. 1977. *The Essential Tension.* University of Chicago Press.

Lakatos, I. 1981. History of science and its rational reconstructions. In *Boston Studies in the Philosophy of Science,* vol. 8. Dordecht.

Lakoff, G. 1986. *Women, Fire, and Dangerous Things: What Categories Reveal about the Mind.* University of Chicago Press.

Latour, B. 1987. *Science in Action.* Harvard University Press.

Leavenworth, I. 1930. *A Methodological Analysis of the Physics of Pascal.* Institute of French Studies, New York.

Leplin, J. 1984. *Scientific Realism.* University of California Press.

Lightman, A., and Gingerich, O. 1992. When do anomalies begin? *Science* 255 : 690–694.

Margolis, H. 1987. *Patterns, Thinking, and Cognition.* University of Chicago Press.

———. 1991. Galileo's *Dialogue* and Tycho's system. *Studies in History and Philosophy of Science* 22 : 259–275.

McNeil, D. 1992. *Hand and Mind.* University of Chicago Press.

Musgrave, R. 1976. Why did oxygen supplant phlogiston? In C. Howson, ed., *Method and Appraisal in the Physical Sciences.* Cambridge University Press.

Neugebauer, O. 1968. On the planetary theory of Copernicus. In *Astronomy and History.* Springer-Verlag.

———. 1975. *A History of Mathematical Astronomy.* Springer-Verlag.

Ore, O. 1953. *Cardano, the Gambling Scholar.* Princeton University Press.

Partington, J. R. 1961. *A History of Chemistry.* Macmillan.

Perrin, N. 1970. Early opposition to the phlogiston theory: Two anonymous attacks. *British Journal for the History of Science* 5 : 128–144.

————. 1987. Revolution or reform: 18th century view of scientific change. *History of Science* 25 : 395–423

————. 1988. The chemical revolution: Shifts in guiding assumptions. In A. Donovan et al., eds., *Scrutinizing Science.* Kluwer Academic Publishers.

Polanyi, M. 1974. *Personal Knowledge.* University of Chicago Press.

Popper, K. 1968. *Conjectures and Refutations.* Harper, Row.

Price, D. J. de S. 1959. Contra-Copernicus. In M. Clagett, ed., *Critical Problems in the History of Science.* University of Wisconsin Press.

Priestley, J. B. 1796. C*onsiderations on the Doctrine of Phlogiston.* Princeton University Press, 1929.

Pylyshyn, Z. 1984. *The Computational Theory of Mind.* MIT Press.

Ragep, J. 1990. Duhem, the Arabs, and the history of cosmology. *Synthèse* 83 : 189–214.

Redondi, P. 1987. *Galileo Heretic.* Princeton University Press.

Rudwick, M. J. S. 1985. *The Great Devonian Controversy.* University of Chicago Press.

Rumelhart, D. 1981. Schemata. In D. Norman, *Perspectives on Cognitive Science.* Erlbaum Associates.

Rumelhart, D., et al. 1986. *Parallel Distributed Processing: Explorations in the Microstructure of Cognition.* MIT Press.

Russell, B. 1956. *Logic and Knowledge.* Macmillan.

Schmitt, C. B. 1951. William Harvey and Renaissance Aristotelianism. Reprinted in *Reappraisals in Renaissance Thought.* Variorum Reprints, 1989.

————. 1969. Experience and experiment: A comparison of Zaberella's view with Galileo's in *De Moto.* Reprinted in *Studies in Renaissance Philosophy and Science.* Variorum Reprints, 1981.

Shapin S., and Schaffer, S. 1985. *Leviathan and the Air Pump.* Princeton University Press.

Sorbière, S. 1664. *Voyage to England.* English trans., 1709.

Swerdlow, N. 1973. The Commentariolus of Copernicus. *Proceedings of the American Philosophical Society* 117 : 428–512.

Thagard, P. 1990. The conceptual structure of the chemical revolution. *Philosophy of Science* 57 : 183–209.

Thoren, V. 1990. *The Lord of Uraniborg.* Cambridge University Press.

Torretti, R. 1990. *Creative Understanding.* University of Chicago Press.

Toulmin, S. 1957. Crucial experiments: Priestley and Lavoisier. *Journal of the History of Ideas* 18 : 205–220.

Tversky, A., and Kahneman, D. 1981. The framing of decisions and the psychology of choice. *Science* 211 : 453–58.

Uleman, J. S., and Bargh, J. A. 1989. *Unintended Thought.* Guilford Press.

Van Helden, A. 1985. *Measuring the Universe.* University of Chicago Press.

————. 1988. Introduction to G. Galileo, *Siderius Nuncius* (1610), trans. A. Van Helden. University of Chicago Press.

Wilson, C. 1975. Rheticus, Ravetz, and the "necessity" of Copernicus' innovation. In R. S. Westman, *The Copernican Achievement.* University of California Press.

————. 1989. Predictive astronomy in the century after Kepler. In R. Taton and C. Wilson, *Planetary Astronomy from the Renaissance to the Rise of Astrophysics, Part A: Tycho Brahe to Newton.* Cambridge University Press.

Wilson, G. [1852] 1975. *Life of Cavendish.* (1852; reprint, Ayer Press).

Ziman, J. 1979. *Reliable Knowledge.* Cambridge University Press.

Index

air pump, Chap. 11 passim; 195; its history, 167–69, 250n5
anomalies, 38–39
anomalous suspension (air pump), 212, 254n7
Apollonius, 123
arguments: as context-dependent, 199; as good, 198–99
Aristarchus, 41, 190
Aristotle, 179, 199, 245n16
Azande logic, 199, 255n10

Bacon, F., 180, 194
Bacon, R., 180
barriers, 2, Chap. 2 passim, 64–67, 69–70, 72, 117, 130, 138–40, 155, 182; vs. gaps, 3, 29–32, 50; as unique, 32, 39–42, 66, 130–31. *See also* facilitators; habits of mind; paradigm shifts
belief matrix, 30
Bellarmine, Cardinal, 195
Bernstein, N., 9–10
black-boxing, 191–92, 253n4
bootstrapping, 197, 198, 243n7
Boyle, R., 5, 146, Chap. 11 passim, 180, 193–95

breakthrough effects, 39, 41–42, 84, 121, 133–34
Branigan, A., 215n5, 220n5
Brouwer, L., 218n5
Bruner, J., 148, 152
Buridan's ass, 148
Buterfield, H., 219n4

caloric vs. mechanical view of heat, 221n5, 223n10, 244n11, 252n3
Cardano, 71, 72, 229n13
Cavendish, 46, 48–54 passim
Chomsky, N., 12
chunking, 139, 150
code-breaking, 252n5
Cohen, I. B., 218n7
coincidence argument against Ptolemy, 117, 124–26
Collins, H., 144, 191, 209
Columbus, 132
comfort and economy, 42, 149 (defined), 155, 156–57, 159–60, 190; and air pump, 172–75
Commentariolus, 135, 234n14; "darkest corner of Europe," 141; *De Revolutionibus,* 114, 135, 234n14; and epicycles,